Fundamentals of Electrotherapy
and
Biomedical Physics

Fundamentals of Electrotherapy and Biomedical Physics

Ashish Kakkad
MPT (Neurological Conditions) MIAP
Assistant Professor
Shri KK Sheth Physiotherapy College
Rajkot, Gujarat, India

JAYPEE BROTHERS MEDICAL PUBLISHERS (P) LTD
New Delhi • London • Philadelphia • Panama

Jaypee Brothers Medical Publishers (P) Ltd

Headquarters
Jaypee Brothers Medical Publishers (P) Ltd
4838/24, Ansari Road, Daryaganj
New Delhi 110 002, India
Phone: +91-11-43574357
Fax: +91-11-43574314
Email: jaypee@jaypeebrothers.com

Overseas Offices

J.P. Medical Ltd
83 Victoria Street, London
SW1H 0HW (UK)
Phone: +44-2031708910
Fax: +02-03-0086180
Email: info@jpmedpub.com

Jaypee Medical Inc
The Bourse
111 South Independence Mall East
Suite 835, Philadelphia, PA 19106, USA
Phone: +1 267-519-9789
Email: joe.rusko@jaypeebrothers.com

Jaypee Brothers Medical Publishers (P) Ltd
Bhotahity, Kathmandu, Nepal
Phone: +977-9741283608
Email: kathmandu@jaypeebrothers.com

Jaypee-Highlights Medical Publishers Inc
City of Knowledge, Bld. 237, Clayton
Panama City, Panama
Phone: +1 507-301-0496
Fax: +1 507-301-0499
Email: cservice@jphmedical.com

Jaypee Brothers Medical Publishers (P) Ltd
17/1-B Babar Road, Block-B, Shaymali
Mohammadpur, Dhaka-1207
Bangladesh
Mobile: +08801912003485
Email: jaypeedhaka@gmail.com

Website: www.jaypeebrothers.com
Website: www.jaypeedigital.com

© 2014, Jaypee Brothers Medical Publishers

The views and opinions expressed in this book are solely those of the original contributor(s)/author(s) and do not necessarily represent those of editor(s) of the book.

All rights reserved. No part of this publication may be reproduced, stored or transmitted in any form or by any means, electronic, mechanical, photocopying, recording or otherwise, without the prior permission in writing of the publishers.

All brand names and product names used in this book are trade names, service marks, trademarks or registered trademarks of their respective owners. The publisher is not associated with any product or vendor mentioned in this book.

Medical knowledge and practice change constantly. This book is designed to provide accurate, authoritative information about the subject matter in question. However, readers are advised to check the most current information available on procedures included and check information from the manufacturer of each product to be administered, to verify the recommended dose, formula, method and duration of administration, adverse effects and contraindications. It is the responsibility of the practitioner to take all appropriate safety precautions. Neither the publisher nor the author(s)/editor(s) assume any liability for any injury and/or damage to persons or property arising from or related to use of material in this book.

This book is sold on the understanding that the publisher is not engaged in providing professional medical services. If such advice or services are required, the services of a competent medical professional should be sought.

Every effort has been made where necessary to contact holders of copyright to obtain permission to reproduce copyright material. If any have been inadvertently overlooked, the publisher will be pleased to make the necessary arrangements at the first opportunity.

Inquiries for bulk sales may be solicited at: jaypee@jaypeebrothers.com

Fundamentals of Electrotherapy and Biomedical Physics

First Edition: **2014**

ISBN 978-93-5090-985-0

Printed at Rajkamal Electric Press, Plot No. 2, Phase-IV, Kundli, Haryana.

Dedicated to
My parents and my teachers

Preface

Fundamentals of Electrotherapy and Biomedical Physics aims to provide basic concepts of electrotherapy to the students. Exercise therapy and electrotherapy are two important components of physiotherapy treatments. By this book, the aim is to provide study materials which will help the students to go into details about all the instruments they are using. Most of the universities have a subject in the first year of bachelor of physiotherapy that teaches about fundamentals of electrotherapy. But, there is no specific book for that. So, by this book, effort is to made to solve the demand of students for the basics of electrotherapy.

The first chapter gives fundamental concepts of circuit, electric charge, electric current, voltage, resistance, etc. to form strong base of further studies for their higher standard in college. Chapter 2 contains ideas about conductors, insulators, and semiconductors, including their examples and uses. Chapter 3 provides the material and detailed knowledge about the construction, mechanism, and types of switches, plugs, sockets and fuses. Chapter 4 explains from how electricity is generated to its use in our home for home appliances, including its generation, distribution, transmission and usage in detail. Chapter 5 contains sources of direct current ranging from simple primary or secondary cell to thermionic valve and choke coil. Chapter 6 will help to form the base of ultrasound and other electric modalities by clearing doubts about skin resistance, conducting medium, and different types of electrodes. Chapter 7 will guide the students how to manage if electric shock or earth shock occurs in physiotherapy department, as causes, types, prevention and treatment of electric shock, and earth shock are explained. Chapter 8 clears the concepts of magnet, magnetic field, and other properties exhibited by magnet. Chapter 9 explains in details about introduction about electromagnetic spectrum and its governing laws such as reflection, refraction, attenuation, cosine law, inverse square law which will help students to fix the placement of different electrotherapeutic instruments for best result. This will also provide base for infrared and ultraviolet rays for future standards of college. This chapter also provides uses of different types of rays used in day-to-day life. Chapter 10 explains about electromagnetic induction in detail that is must to know before understanding the transformer. Chapter 11 elaborates transformer with its types such as step-up, step-down, etc. for good understanding of voltage regulations. Chapter 12 explains the details for transistor with its types and mechanism of work. Chapter

13 provides base for understanding about capacitor, including its construction, types and uses which will help the reader learn shortwave diathermy easily for getting maximum output from machine. Chapter 14 gives detailed knowledge about pain and its regulation so that the reader will be able to understand that whatever electrotherapy modality he/she is using for pain relief that works on which level and mechanism.

The effort is also made to provide all study materials with references from different books of electrotherapy. To get recent advances of all covered topics reference websites are also mentioned at the end of the chapter.

For freshers, most of the time, they have reading materials; but, they are confused how to face this in examinations. To solve this to level best, effort is made to provide multiple choice questions (MCQs) with answers.

All the readers are requested to provide their suggestions regarding this book so that next time precaution can be taken. Feel free to contact on *kakkadashish@yahoo.co.in*. The author hopes to get best response from the readers.

Ashish Kakkad

Contents

1. **Fundamentals of Electric Circuit**   1

   Theories of Electricity   1
   Electric Charge   2
   Electric Current   3
   Electric Field   3
   Voltage or Potential Drop or Electric Tension or Potential Difference or Electromotive Force or Electrical Potential Difference   4
   Electric Resistance   5
   Ohm's Law   6
   Grouping of Resistance   6
   Electric Conductance   7
   Electrical Energy or Electric Potential Energy or Electrostatic Potential Energy   8
   Electric Power   8
   Joule's Law or Joule's Effect or Joule–Lenz Law   9
   Coulomb's Law or Coulomb's Inverse-Square Law   9
   Rheostat   10
   Ammeter   11
   Voltmeter   12

2. **Conductors, Semiconductors and Insulators**   16

   Conductors   16
   Insulators   16
   Semiconductors   16

3. **Power Plugs, Sockets, Switches and Fuses**   20

   Power Plugs and Sockets   20
   Switches   21
   Fuse   21

4. **Electricity Types, Generation Transmission, Distribution and Usage**   24

   Electricity Types   24
   Parameters of Alternating Current   25
   Electricity Generation   27
   Electricity Transmission and Distribution   29
   Mains Supply   30
   Earthing   31

5. Sources of Direct Current — 34

Cell 34
Thermionic Valves 35
Diode 35
Triode 37
Rectifier 37
Choke Coil 38

6. Skin Resistance, Electrodes and Gel — 41

Skin Resistance 41
Electrodes 42
Electrode Gel 45

7. Electric Shock — 48

Earth Shock 52

8. Magnet and its Properties — 56

Geometry of Magnet 56
Terminologies 57
Types 57
Othertypes of Magnets 58
Shapes 59
Magnetic Effect of Electric Current 59
Atomic/Molecular Theory of Magnetism 59
Properties of a Magnet 60
Properties of Magnetic Lines of Force 60

9. Electromagnetic Spectrum—its Uses and Governing Laws — 64

Terminologies 64
History 65
Types 66
Reflection 67
Refraction 67
Snell's Law or Snell–Descartes Law or Law of Refraction 68
Absorption 69
Inverse Square Law 69
Attenuation or Extinction 70
Cosine Law or Lambert-Cosine Law 72
Grothus Law or Grotthus Draper Law 72
Uses of Electromagnetic Waves 73

10. Electromagnetic Induction — 79

Self-induction 80
Mutual Induction 80
Inductive Reactance 81

Faraday's Law   81
Lenz's Law   82
Eddy Current or Foucault Current   82

11. Transformer   86

12. Transistor   91

13. Capacitor or Condenser   95

Capacitive Reactance   100

14. Physiology of Pain   102

Index   *109*

Fundamentals of Electric Circuit 1

THEORIES OF ELECTRICITY

Electronics

Electronics is the field of manipulating electrical currents and voltages using passive and active components that are connected together to create circuits. Electronic circuits range from a simple load resistor that converts a current to a voltage, to computer central processing units that can contain millions of transistors. Electronic devices operate by the movement of electrons through conductors.

Active Components

They change their resistance or impedance when varying voltages are applied to them and as a result can amplify, rectify, modify or distort alternating current waveforms. For example vacuum tubes, diodes, transistors.

Passive Components

They don't alter their resistance, impedance or reactance when alternating currents are applied to them. They normally don't distort waveforms. For example Resistors, inductors, transformers and capacitors.

Matter and Electricity

All matter consists of molecules. A molecule can be defined as the smallest particle, which shows all the characteristics of a particular matter. For example, molecule of water is obtained by dividing a drop of water again and again until it can be divided and still is water. Further division of this water molecule will yield three particles which are not water. Molecule of water contains two atoms of hydrogen and one atom of oxygen.

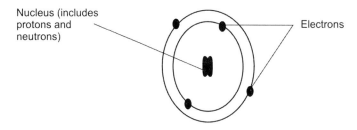

Fig. 1.1: Atom with protons, neutrons and electrons

Chemical combination of different atoms makes a molecule. An atom can be further divided into three particles known as protons, electrons and neutrons (Fig. 1.1). These particles cannot be divided further.

ELECTRIC CHARGE

As we know, protons and electrons are the particles possessing electrical properties whereas neutron is electrically neutral. Charge is amount of electrons or protons.

Electrons are the negatively charged particles, which revolve around the positively charged protons which are located in the nucleus of an atom along with neutrons. Proton is about 1800 times heavier than electron. There is always attraction between unlike charges. Because electron is much lighter than proton, hence it is pulled towards the proton. If the force of attraction is enough, then the electron comes too closer to the proton and both the particles together form a neutral particle to be known as neutron.

The electrical charge of an electron can be explained with the help of an imagination that there exist lines of forces, which are outward pointing. Though the size and weight of electron and proton varies significantly, the negative field of an electron is just as strong as the positive field of a proton. Though it is small physically, the field near the electron is quite strong. The strength of the field varies inversely with the distance squared.

Though electrons and protons have different kind of charge in them, both have charges of equal magnitude. An electron (negatively charged) repels another electron, while a proton (positively charged) repels another proton. But the proton and electron have attracting force between them. As electron is lighter, electron will be attracted towards the proton. So the basic physical law states that:

"Like charges repel each other; unlike charges attract each other" (Fig. 1.2).

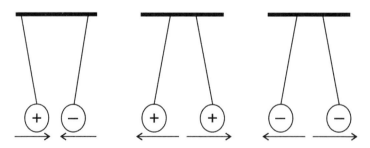

Fig. 1.2: Attraction between like charges and repulsion between like charges

Unit

- Coulomb (C)
- One coulomb is equivalent to 6×10^{18} electrons

Symbol

- q, –q or Q.

ELECTRIC CURRENT

Electric current is the rate of flow of electric charge. Electric current is a flow of electric charge, mainly the electron, from one point in unit time, through a medium such as wire in unit time. It can also be carried by ions.

Direction

Before some years, it was believed that flow of positive charge occurs from anode to cathode so direction of conventional current was given from anode to cathode. In later years, it was found that electric current was due to negatively charged particles electrons. Electrons move from cathode to anode. But still we use to indicate direction of current from anode to cathode only.

Formula

- $I = Q / t$
- Where I is amount of current (in ampere)
- Q is electric charge (in coulomb) passing through the cross-sectional area
- t is time to pass from given surface area (in seconds).

Measurement

By ammeter

Unit

- Ampere (A)
- 1 ampere means one coulomb charge flowing through cross-sectional area of conductor in one second.

Symbol

I

Most commonly in liquids, electric currents are because of movement of electrolytes. Electrolytes are electrically charged particles (ions). For example if a current is passed in a solution of sodium chloride, the sodium ions move towards the negative electrode, while the chloride ions move towards the positive electrode.

In air and other ordinary gases electrical conductivity is low due to few mobile ions. Mostly air and gases are insulators. Vacuum contains no charged particles so it behaves as an insulator.

ELECTRIC FIELD

Michael Faraday gave the concept of electric field. It is the region around the charged particle or timely changing magnetic field, where if another charged particle comes; it will feel either attracting or repelling force depending on its polarity (Fig. 1.3).

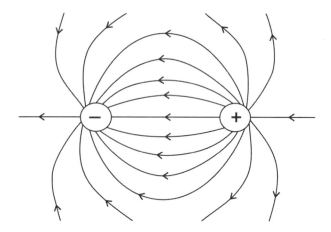

Fig. 1.3: Electric field and electric lines of force

Characteristics of Electric Lines of Force

- Electric lines of force are just imaginary lines which show spreading of electric field.
- The direction of electric lines of force is from positive charge to negative charge.
- Electric lines of force reduced longitudinally because of attraction between unlike charges.
- Electric lines of force expand laterally because of repulsion between like charges.
- Electric lines of forces do not intersect each other.
- Electric lines of force enter or leave any conductor at right angle to surface of conductor.
- Electric lines of force are open loops unlike magnetic lines of force.
- The concentration of electric lines of force will increase if radius of conductor in which it enters is less.

VOLTAGE OR POTENTIAL DROP OR ELECTRIC TENSION OR POTENTIAL DIFFERENCE OR ELECTROMOTIVE FORCE OR ELECTRICAL POTENTIAL DIFFERENCE

Electric charge is location dependent property of circuit. In two different points of circuit, amount of electric charge will be different. In simpler words, we can assume that on particular location, electric charge has its potential energy (Fig. 1.4).

So just for example, there are two points A and B in the circuit. Point A is of high potential energy means containing high amount electric charge as compared to point B. So if charge has to be moved from A to B, it will be without energy. But if electric charge is to be moved from B to A, work has to be done. Potential difference is the difference in electric potential energy per unit charge between two points of circuit. A voltage may represent either a source of energy, or it may represent lost or stored energy.

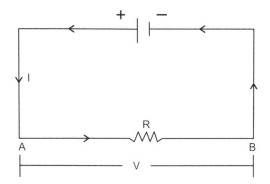

Fig. 1.4: Electric circuit where I is electric current, R is resistance, V is potential difference

Production

Voltage can be caused:
- By static electric fields
- By electric current through a magnetic field
- By time-varying magnetic fields
- A combination of all three.

Measurement

- By voltmeter.

Unit

- Volts (V), or Joules per coulomb
- One volt is the potential difference that requires one Joule of energy to move one coulomb of charge.

Symbol

- ΔV.

ELECTRIC RESISTANCE

The electrical resistance of circuit is the opposition to the passage of an electric current through conductor. More the electric resistance, less the current passes.

The resistance of an object is defined as the ratio of voltage to electric current through the circuit.

Formula

- $R = V / I$
- Where R is electric resistance of object (in ohm)

- V is potential difference between two points of circuit (in volt)
- I is current passing from conductor (in ampere).

Units

- Ohm (Ω).

Symbol

- R.

OHM'S LAW

It states that the current through a conductor between two points is directly proportional to the potential difference across the two points and inversely proportional to resistance. The law was named after the German Physicist Georg Ohm, and Published in 1827.

Formula

- I = V / R.

Where I is the current through the conductor in defined cross-sectional area (in ampere), V is the potential difference between two points of circuit (in volt) and R is the constant electric resistance of conductor (in ohm).

Here resistance R is independent of current, while voltage and current are interdependent.

GROUPING OF RESISTANCE

Grouping of resistance can be done in two types:
- Resistance in series (Fig. 1.5)
- Resistance in parallel (Fig.1.6).

Resistance in Series

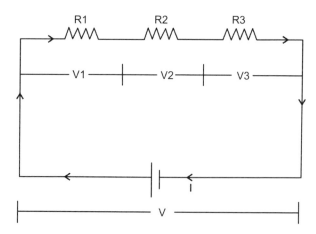

Fig. 1.5: Resistance in series

From Figure, 1.5.
- $V = V_1 + V_2 + V_3$ ....(1)

According to Ohm's law,
- $V = IR$

From formula (1)
- $V_1 + V_2 + V_3 = IR$
- $V_1/I + V_2/I + V_3/I = R$
- $R_1 + R_2 + R_3 = R$
- In general, $R = R_1 + R_2 + R_3 + \text{---} + R_n$
- Where n is number of resistance in series.

Resistance in Parallel

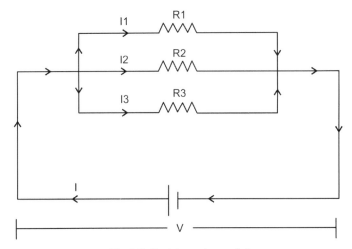

Fig.1.6: Resistance in parallel

From Figure, 1.6.
- $I = I_1 + I_2 + I_3$ ....(2)

According to Ohm's law,
- $V = IR$

From formula (2),
- $V = (I_1 + I_2 + I_3) R$
- $1/R = (I_1 + I_2 + I_3) / V$
- $1/R = I_1/V + I_2/V + I_3/V$
- $1/R = 1/R_1 + 1/R_2 + 1/R_3$
- In general, $1/R = 1/R_1 + 1/R_2 + 1/R_3 + \text{---} + R_n$

Where, n is number of resistances in parallel.

ELECTRIC CONDUCTANCE

The inverse of electric resistance is called electrical conductance. More the conductivity, more the current passes.

The conductance of an object is defined as the ratio of electric current through it to voltage.

Formula

- G = I / V
- Where G is electric conductance of object (in mho)
- V is potential difference between two points of circuit (in potential difference)
- I is current passing from conductor (in ampere).

Unit

- Siemens (S) or mho (℧).

Symbol

- G.

ELECTRICAL ENERGY OR ELECTRIC POTENTIAL ENERGY OR ELECTROSTATIC POTENTIAL ENERGY

Electrical energy is the energy gained from any form of electricity.
It can be related to following:
- The energy stored in an electric field
- The potential energy of a charged particle in an electric field
- The energy provided by electricity.

This form of energy is not well known as other types of energy like kinetic energy, potential energy, etc. Electrical energy is the presence and flow of an electric charge. The energy portion of electricity is found in a variety of phenomena such as static electricity, electromagnetic fields and lightning.

We are not getting electrical energy directly in ready to use form. Electrical energy can be gained by transformation of other energies into it. For example, by rotating turbines from waterfall, kinetic energy can be converted into electrical energy. From battery, chemical energy can be converted into electrical energy. From sunlight, solar energy can be converted into electrical energy and so on.

Unit

- Joule (J).

ELECTRIC POWER

Electric power is the amount of work done by an electric current in a unit time. Electric power is the rate at which electric energy is transferred by an electric circuit. When a current flows in a circuit against resistance, work is done. Instruments can be made those convert this work into heat (electric heaters), light (light bulbs and neon lamps), or motion, i.e. kinetic energy (electric motors).

Formula

- $P = IV = I^2R = \dfrac{V^2}{R}$

(Replace I or V in P = IV according to Ohm's law V = IR),
Where, P is power (in watts),
- I is current (in ampere)
- V is potential difference (in volt)
- R is electrical resistance (in ohm).

Unit

- Watt (W) or joule per second.

Symbol

- P.

JOULE'S LAW OR JOULE'S EFFECT OR JOULE–LENZ LAW

Joule's laws are a pair of laws about amount of heat production in electric circuit and internal energy of ideal gas. They are named after James Prescott Joule.

First Law

When electric current passes through any conductor, heat energy is produced in it. It is due to the collision of electrons with the atoms. Joule gave relationship between heat produced in the conductor, intensity of current, resistance of conductor and time for current to pass. It was given by Joule in 1840. This heating effect is known as Joule heating. It is also called the Joule–Lenz law since it was later independently discovered by Heinrich Lenz.

Formula

- $Q = I^2 Rt$

Where Q is the heat generated by a constant current (joule)
I is intensity of electric current (ampere)
- R is resistance of conductor (ohm)
- t is time for current to pass in conductor (second).

Second Law

It states that the internal energy of an ideal gas is independent of its volume and pressure, depending only on its temperature.

COULOMB'S LAW OR COULOMB'S INVERSE-SQUARE LAW

This law describes interaction between electrically charged particles. It was first published in 1785 by Charles Augustin de Coulomb and was essential to the development of the theory of electromagnetism.

Force between two charges is directly proportional to multiplication of two charges values and inversely proportional to square of distance between two charges.

Formula

- $F = kq_1q_2/r^2$
- Where F is force between two electric charges (in newton)
- k is coulomb force constant or coulomb constant
- q1 and q2 are two different electric charges (in coulomb)
- r is distance between two electric charges (in meter).

RHEOSTAT

The most common way to vary the resistance in a circuit is to use a rheostat.

Construction

t is made up of coil of a resistance wire wound around a semicircular insulator, with the wiper sliding from one turn of the wire to the next. Each turn of wire is insulated from successive turn of wire.

Variable Resistance or Series Rheostat (Fig. 1.7)

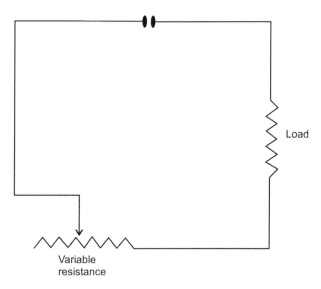

Fig.1.7: Series rheostat

In this wires are in arranged series, so more number of wires when involved in circuit, amount of resistance increases and amount of current decreases. When number of wires in for offering resistance in circuit decreases, amount of current increases. These types of apparatus are not advisable for patients to apply current to their body as current cannot be made zero in these apparatus. They can be used in other apparatus where there is no placement on patients' body.

For example, paraffin wax bath where no contact of machine with patients is needed.

Potentiometer or Shunt Rheostat (Fig. 1.8)

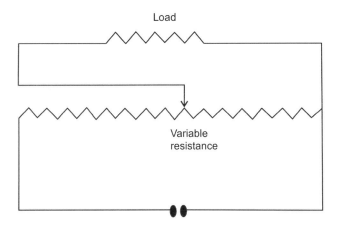

Fig 1.8: Shunt rheostat

It is wired across a course of potential difference and any other circuit has to be taken off in parallel to it. This acts by altering the potential difference between the ends of circuit. When potential difference is more, amount of current will be produced more. When potential difference is less, amount of current will be produced less. This type of apparatus used where direct placement on patient's body as current can be made from zero upto maximum.

Other Uses

Wire-wound rheostats made with ratings up to several thousand watts are used in applications such as DC motor drives, electric welding controls, or in the controls for generators.

Symbol

- ⸺⸺ or ⸺⋀⋀⸺

AMMETER

An ammeter (Fig. 1.9) is the device used to measure the electric current in a circuit. Electric currents are measured in amperes the name given is ammeter.

Most of them are used to measure electric current in milliampere or microampere so they are also known as milliammeters or microammeters. By the late 19th century, better instruments have been designed for more precised measurement.

Fig 1.9: Ammeter

Symbol

-

VOLTMETER (FIG. 1.10)

It is an instrument used for measuring electrical potential difference between two points in an electric circuit.

Fig. 1.10: Voltmeter

There are two types of voltmeters:
- Analog voltmeter
- Digital voltmeter.

Analog voltmeters move a pointer in proportion to the voltage of the circuit, while digital voltmeters give a numerical value in numbers of voltage by use of an analog to digital converter. The first digital voltmeter was invented and produced by Andrew Kay in 1954.

Voltmeter accuracy is affected by following factors:
- Temperature
- Supply voltage variations.

An ideal voltmeter has infinite resistance. It is not possible to make a voltmeter with infinite resistance so a well-designed voltmeter should contain high resistance.

Symbol

- —(V)—

MULTIPLE CHOICE QUESTIONS

1. Electric field lines pass…
   a) From positive charge to negative charge
   b) From positive charge to positive charge
   c) From negative charge to positive charge
   d) From negative charge to negative charge
2. Comment about statement "electric field lines do not cross each other."
   a) True
   b) False
   c) Cannot be commented
   d) Depends on polarity of charge
3. The electric lines of force enter into any surface at
   a) 180° to surface
   b) 90° to surface
   c) 45° to surface
   d) 0° to surface
4. As the radius of curvature of surface, where lines of force enters increases …
   a) Concentration of lines of force increases
   b) Concentration of lines of force is not affected by radius of curvature of surface
   c) Concentration of lines of force decreases
   d) None of the above
5. Following is not unit of potential difference…
   a) Volt
   b) Joules per coulomb
   c) Both of the above
   d) None of the above
6. Potential difference is measured with…
   a) Ammeter
   b) Voltmeter
   c) Capacitor
   d) Transistor
7. Identify one that is not matching
   a) Potential difference

b) Electromotive force
c) Volt
d) Potential drop
8. Force between two charges "q1" and "q2" separated by distance "r" will be...
   a) Directly proportional to product of q1, q2 and r^2
   b) Directly proportional to product of q1 and q2 and inversely proportional to r
   c) Directly proportional to product of q1 and q2 and inversely proportional to r^2
   d) Inversely proportional to product of q1 and q2 and directly proportional to r^2
9. The function of the rheostat is ...
   a) To alter the voltage in the circuit
   b) To alter the electric current in the circuit
   c) To alter the resistance of the circuit
   d) To alter the power of the circuit
10. Following type of rheostat is used where machine is used for application of current to patient's body directly
    a) Series rheostat
    b) Shunt rheostat
    c) Both of the above
    d) None of the above
11. Pick up correct statement
    a) An ideal voltmeter has infinite resistance
    b) An ideal voltmeter has zero resistance
    c) An ideal voltmeter has ability to alter voltage in the circuit
    d) Ideal voltmeter accuracy is not affected by external factors
12. Function of the ammeter is...
    a) Measurement of resistance of circuit
    b) Measurement of voltage between two points of circuit
    c) Measurement of electric current in the circuit
    d) Measurement of electric power in the circuit

Answers:

1—a	5—d	9—c
2—a	6—b	10—b
3—b	7—c	11—a
4—c	8—c	12—c

Bibliography:
1. Clayton's Electrotherapy. AITBS Publishers, 2000.
2. Kr Khokhar. Helpline Electrotherapy for Physiotherapists, 2nd ed. Bharat Bharti Prakashana Co., 2005.
3. Singh J. Textbook of Electrotherapy, 1st ed. Jaypee Brothers Medical Publishers (P) Ltd. 2007.

Reference websites:
1. http://en.wikipedia.org
2. http://ww.answers.yahoo.com
3. http://www.circuitstoday.com
4. http://www.citycollegiate.com
5. http://www.ee.duke.edu
6. http://www.hyperphysics.phy-astr.gsu.cdu
7. http://www.thebigger.com
8. http://www.wisegeek.com

Conductors, Semiconductors and Insulators | 2

The electrons moving around the nucleus can be moved from an atom to another atom and from object to object. These electrons will move depending on whether the material is a conductor or an insulator. Some materials fall between these two categories, they are known as semiconductors.

CONDUCTORS

Some of the electrons in it are held loosely by the atom. Such electrons move freely from atom to atom within the material. So these types of materials will allow the electric current to pass. Copper has very high conductivity. Silver has more conductivity than copper, but due its higher cost, not used much. Aluminum has only approximate 60% conductivity than that of copper, but because of its low cost, its use is prevalent.

INSULATORS

In this, the electrons are held tightly to the atom and are not able to move freely within the material. So these types of materials either will not allow or will allow to very less extent the electric current to pass. Glass, papers, plastic are most commonly used insulators.

SEMICONDUCTORS

It has electrical conductivity intermediate between that of a conductor and an insulator. Metals are good conductors. Metal conductivity decreases with temperature increase because it will disturb the free motion of electrons. Insulators are very poor conductors of electricity. Insulator conductivity increases with temperature. Semiconductors, on the other hand, have an intermediate level of electric conductivity when compared to metals and insulators.

Semiconductors are insulators at low temperature as temperature and resistance of semiconductor are inversely proportional. It should be noted, that the negative charge of the electrons is balanced by an equivalent positive charge in the center of the impurity atoms. Therefore, the net electrical charge of the semiconductor material is not changed.

Uses

Semiconductors are used in modern electronics, e.g. radio, computers, and telephones.

Types

They are of two types:
- Intrinsic semiconductor
- Extrinsic semiconductor.

Intrinsic semiconductor

An intrinsic semiconductor is made up ideally of one pure element, typically silicon. At room temperature, the conductivity of intrinsic semiconductors is relatively low.

Extrinsic semiconductor

Conductivity of intrinsic semiconductor is greatly enhanced by a process called doping, in which other elements are added to the intrinsic crystal in very small amounts to create what is called extrinsic semiconductor.

Extrinsic semiconductors have subtypes

- N-type (N for negative)
- P-type (P for positive).
- N-type semiconductors have more number of electrons
- P-type semiconductors have less number of electrons.

If above mentioned body semiconductors are fused together current will flow from N to P side only. By this property only, they are used as valve.

P-type semiconductor (Fig. 2.1)

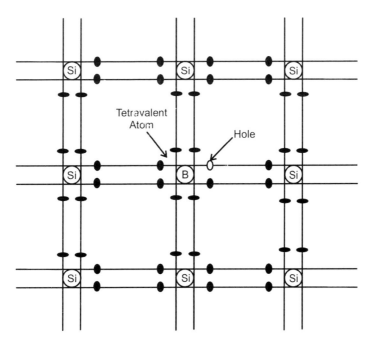

Fig. 2.1: P-type semiconductor

It is obtained by carrying out a process of doping by adding a certain type of atoms (acceptors) to the semiconductor in order to increase the number of free charge carriers (in this case positive holes). In the case of silicon, a trivalent atom like indium or boron is added into the crystal lattice. Silicon has four valence electrons. Out of these four electrons, three valence electrons will make covalent bond with three valence electrons of trivalent atoms like indium or boron. In structure lattice, one hole is formed because of lack of one electron. These holes will be working as positive charge so these types of semiconductors are also known as P-type semiconductors. In these semiconductors, added materials are accepting electron from Silicon so they are also known as acceptor-type semiconductors.

N-type semiconductors (Fig. 2.2)

They are a type of extrinsic semiconductor where the dopant atoms (donors) are capable of providing extra conduction electrons to the host material. This creates an excess of negative (N-type) electron charge carriers. For example, in tetravalent atoms of silicon, if impurities like pentavalent atoms of arsenic or phosphorus, etc. are added, four valence electrons of Silicon will make covalent bonds with four valence electrons of pentavalent electrons. Still fifth valence electrons of these pentavalent atoms will be free to move. As flow of current is due to this free electron, these types of semiconductors are known as N-type semiconductors. In these types of semiconductors electrons are donated by added materials (dopants), so they are also known as donor-type semiconductors.

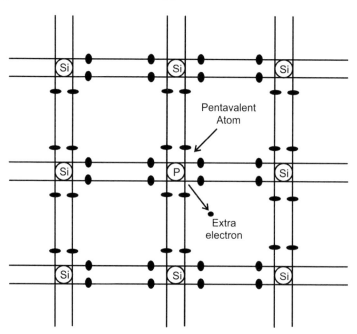

Fig. 2.2: N-type semiconductor

Conductors, Semiconductors and Insulators

MULTIPLE CHOICE QUESTIONS

1. Following is not conductor.
   a) Copper
   b) Aluminum
   c) Silver
   d) Plastic
2. In P-type semiconductor, flow of electricity is because motion of…
   a) Electron
   b) Proton
   c) Neutron
   d) Positive hole
3. Following is intrinsic semiconductor
   a) Phosphorus
   b) Boron
   c) Indium
   d) Silicon

Answers:
1—d
2—d
3—d

Bibliography:
1. Singh J. Textbook of Electrotherapy, 1st ed. Jaypee Brothers Medical Publishers (P) Ltd. 2007.

Reference website:
1. http://en.wikipedia.org

Power Plugs, Sockets, Switches and Fuses | 3

POWER PLUGS AND SOCKETS (FIGS 3.1 A AND B)

AC power plugs and sockets are devices for irremovably connecting electrically operated devices to the power supply. An electric plug is a male electrical connector with contact prongs to connect mechanically and electrically to slots in the matching female socket. Wall sockets (sometimes also known as power points, power sockets, electric receptacles, plug sockets, electrical outlets or just sockets) are female electrical connectors that have slots or holes which accept and deliver current to the prongs of inserted plugs. To reduce the risk of injury or death by electric shock, some plug and socket systems incorporate various safety features. Sockets are designed to accept only matching plugs and reject all others.

Fig 3.1A: Socket   **Fig 3.1B:** Plug

Each receptacle has two or three wired projections. The projections may be steel or brass, and may be plated with zinc, tin, or nickel. The live contact carries current from the source to the load. The neutral contact returns current to the source. Many receptacles and plugs also include a third contact for a connection to earth ground, intended to protect against insulation failure of the connected device. A common approach is for electrical sockets to have three holes, which can accommodate either 3-pin earthed or 2-pin nonearthed plugs. Electrical plugs and their sockets differ by country in shape, size and type of connectors. The type used in each country is set by national standards. For direct current and alternating current, dimension of plug and socket will vary.

SWITCHES (FIG. 3.2)

The current turns off and on by means of switch. Switches vary in types according to the current that is to be passed through them. The ones commonly used in house and physiotherapy departments consist of two metal blades which fit into metal sockets. Each set of contacts can be in one of two states: either "closed" meaning the contacts are touching and electricity can flow between them, or "open", meaning the contacts are separated and the switch is not conducting. The principle is that when the switch is on, the blades are gripped in the sockets and the circuit is completed. When the circuit is broken, a spring ensures the sudden separation of the sockets and blades.

There is a switch for each light and power point and it is most satisfactory if this breaks both wires and circuit. When it is not so the connection of live wire can be made even when the switch is turned off and danger of earth shock exists.

Fig 3.2: Switches

Operation of switches varies with its type. Most of switches are manually operated. They can be in two conditions: like on or off. Examples are switches used for lights and fans in buildings. Some of the switches are in same position during on and off conditions. Once you press, it will be on and second time when you press, it will be off. Example includes switch of mobile of computer, etc. Some switches are automatic. They are regulated by another parameters like temperature, pressure, etc. For example, in refrigerator, air conditioning system after reaching trigger point, switch will be off automatically.

FUSE (FIG. 3.3)

A fuse is designed to be a weak point in a circuit which blows if current of too great intensity is passed. It is a type of low resistance resistor that acts as a sacrificial device to provide protection from excessive current. It is considered security guard of machine so connected with circuit of machine in series. Its essential component is a metal wire or strip that melts when too much current flows, which interrupts the circuit in which it is connected. Short circuit, overloading, mismatched loads or device failure are the prime reasons for excessive current.

It consists of short length of wire of low melting point and if the current passing through it exceeds a certain value the generated heat melts the wire. This breaks the circuit and prevents the further current flow and possible damage to another part of the wiring or overheating which might cause fire and gives warning of the defect which caused extreme current. The fuse is placed at the point where wire from mains enters and where the heat generated can cause no damage. It is an essential safety device in any wiring system.

Fig 3.3: Fuse

The most common type of fuse is the cartridge in which the fusible element is made of silver and runs between metal caps through the tube of glass. It is held in position by metal contact clips. The whole tube is replaced when necessary.

In a physiotherapy department fuses should be included in the circuit of each piece of apparatus used for the treatment of the patients and in addition to those in department wiring.

MULTIPLE CHOICE QUESTIONS

1. Following is essential safety device in any electrical device
   a) Electrical fuse
   b) Electrical switch
   c) Power socket
   d) None of the above
2. Following is not component of switch
   a) Spring
   b) Metal blade
   c) Metal socket
   d) Silver wire
3. Following are synonyms for sockets
   a) Power points
   b) Electric receptacles
   c) Electrical outlets
   d) All of the above

Answers:

1—a
2—d
3—d

Bibliography:

1. Clayton's Electrotherapy. AITBS Publishers, 2000.
2. Kr Khokhar. Helpline Electrotherapy for Physiotherapists, 2nd ed. Bharat Bharti Prakashana & Co., 2005.

Reference websites:

1. http://en.wikipedia.org
2. http://www.allaboutcircuits.com
3. http://www.orbit-computer-solutions.com

Electricity Types, Generation Transmission, Distribution and Usage

4

ELECTRICITY TYPES

There are two types of electricity:
- Static electricity
- Current electricity.

Static electricity is made by rubbing together two or more objects and making friction while current electricity is the flow of electric charge across an electrical field.

Static Electricity

Static electricity is when electrical charges build–up on the surface of a material. It is usually caused by rubbing materials together. The result of a build-up of static electricity is that objects may be attracted to each other or may even cause a spark to jump from one to the other. For example rub a balloon on wool and hold it upto the wall.

Before rubbing, like all materials, the balloons and the wool sweater have a neutral charge. This is because they each have an equal number of positively charged subatomic particles (protons) and negatively charged subatomic particles (electrons). When you rub the balloon with the wool sweater, electrons are transferred from the wool to the rubber because of differences in the attraction of the two materials for electrons. The balloon becomes negatively charged because it gains electrons from the wool, and the wool becomes positively charged because it loses electrons.

Current Electricity

Current is the rate of flow of electrons. It is produced by moving electrons and it is measured in amperes. Unlike static electricity, current electricity must flow through a conductor, usually copper wire. Current with electricity is just like current when you think of a river. The river flows from one spot to another, and the speed it moves is the speed of the current. With electricity, current is a measure of the amount of energy transferred over a period of time. That energy is called a flow of electrons. One of the results of current is the heating of the conductor. When an electric stove heats up, it's because of the flow of current.

There are different sources of current electricity including the chemical reactions taking place in a battery. The most common source is the generator. A simple generator produces electricity when a coil of copper

Electricity Types, Generation, Transmission, Distribution and Usage

turns inside a magnetic field. In a power plant, electromagnets spinning inside many coils of copper wire generate vast quantities of current electricity.

There are two main kinds of electric current.
- Direct current (DC) (Fig. 4.1A)
- Alternating current (AC) (Fig. 4.1B).

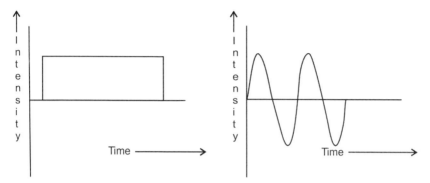

Figs 4.1A and B: Direct current and alternating current

It's easy to remember. Direct current is the current which will not change its direction. In direct current polarity of both electrodes are fixed. Alternating current will change its direction continuously. In alternating current polarity of both electrodes will change continuously.

Heat is developed in all type of electrical circuits due to the flow of electric current. The amplitude of the DC being constant produces more heat in a circuit compared to the heat produced by an AC. In long distance transmission lines, large amount of power will be wasted if DC is used but that can be minimized by the use of AC.

PARAMETERS OF ALTERNATING CURRENT (FIG. 4.2)

Mostly alternating current is applied to patient for therapeutic purpose except in Iontophoresis in which ions are introduced into the human body by direct current. Following are the parameters of alternating current.
- *Continuous or uninterrupted alternating current*: It is bidirectional flow of alternating current without interruption.
- *Pulsed or pulsatile or interrupted alternating current*: It is flow of alternating current with periodic ceases for definite period of time.
- *Monophasic pulsed current*: In this type, flow of alternating current is unidirectional.
- *Biphasic pulsed current*: In this type, flow of alternating current is bi-directional.
- *Pulse and phase*: It is an isolated electrical event separated by definite time from next electrical event. Pulse includes flow of electric current on both directions. Phase is a part of pulse which indicates flow of electric current in only one direction. One pulse has either one or two phases.

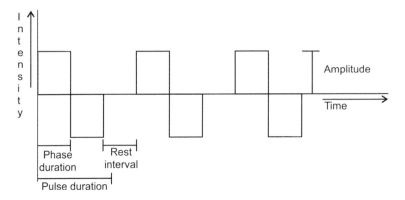

Fig. 4.2: Parameters of alternating current

- *Pulse duration*: It is the time between onset of pulse and termination of same pulse to complete one complete cycle of bidirectional flow of alternating current.
- *Phase duration*: It is the time between onset of phase and termination of same phase to complete unidirectional flow of alternating current.
- *Intensity or amplitude or magnitude*: It is the amount of maximum current reached in one phase.
- *Rise time*: It is the time taken by phase to rise from zero intensity to maximum intensity.
- *Decay time*: It is the time taken by pulse to decrease from maximum intensity to zero intensity.
- *Interpulse interval*: It is time between termination of one pulse and onset of next successive pulse.
- *Waveform (Fig.4.3)*: It is geometric shape of the pulse or phase as they appear on the graph of current or voltage versus time graph. It may be of following types, mainly:

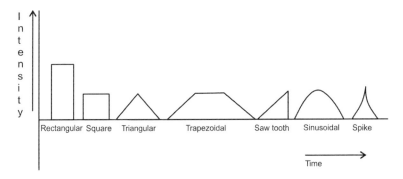

Fig. 4.3: Waveforms

- Rectangular
- Square
- Triangular

- Trapezoidal
- Sawtooth
- Sinusoidal
- Spike.
• *Frequency*: It is number of complete pulses passes through a fixed point in space in a unit time.
  – *Unit*: For general purpose: Hertz (Hz)
  – *For continuous alternating current*: Cycles per second (cps)
  – *For pulsed alternating current*: Pulses per second (pps)
  – *Classification of current according to frequency*:
    Low frequency current: Up to 1,000 Hz
  Medium frequency current: 1,000 Hz to 10,000 Hz
  High frequency current: More than 10,000 Hz.

ELECTRICITY GENERATION

Basically India is agriculture based country. From last few years industrial growth has also increased. In both sectors electricity is required. So electrical energy is essential for the growth of our country. The availability of good power supply provides benefits to almost every sector of a nation. So the Government gives priority to power sectors for development because without electricity industrial or agriculture growth is not possible.

Power cuts were very common in India as compared to developed country. Now in the 21st century finds a huge number of electric power plants are located across India. India has sufficient technology and expertise to generate electricity through the use of —
- Coal power
- Wind power
- Water power
- Nuclear power.

Coal-based plants are the main source of fuel for the electric plants in India. These are run by coal supplies from mines situated in several states of the country. Since 1947, the Government of India has given special attention to the exploitation of coal. Coal is one of the prominent natural resources in India. States like Madhya Pradesh and West Bengal are rich in coal. These take care that the thermal power plants in India are not short of coal supply. The first demonstration of electric light in Calcutta was conducted in 1879.

Power development in India was first started in 1897 in Darjeeling. Now India has been one of the fastest growing markets for new electricity generation capacity. In India, government and private both sectors are working but private sectors are growing at faster rate.

In general, electricity can be generated through following ways:
- Static electricity, from the physical separation and transport of charge (examples: triboelectric effect and lightning).
- Electromagnetic induction, where an electrical generator or dynamo-transforms kinetic energy into electricity.

- Electrochemistry, the direct transformation of chemical energy into electricity, as in a battery, fuel cell or nerve impulse.
- Photoelectric effect, the transformation of light energy into electrical energy, as in solar cell.
- Thermoelectric effect, direct conversion of temperature differences to electricity, as in thermocouples.
- Piezoelectric effect, from the mechanical strain of electrically anisotropic molecules or crystals.
- Nuclear transformation, the creation and acceleration of charged particles (examples: Betavoltaics or alpha particle emission).

Details of different electricity generation of are given here:

Thermal Power

Thermal power plants convert energy rich fuel into electricity and heat. Possible fuels include coal, natural gas, petroleum products, agricultural waste and domestic trash / waste. Other sources of fuel include landfill gas and biogases. In some plants, renewal fuels such as biogas are cofired with coal. The state of Maharashtra is the largest producer of thermal power in the country.

Hydropower

Artificial dams or natural falls of water are used to rotate turbines implanted. This will produce electrical energy.

Nuclear Power

India's nuclear power plant development began in 1964 after agreement with United States. India has nuclear power plants operating in the following states: Maharashtra, Gujarat, Rajasthan, Uttar Pradesh, Tamil Nadu and Karnataka. India has also some amount of uranium from mine. India's share of nuclear power plant generation capacity is just 1.2% of worldwide nuclear power production capacity.

Other Renewable Energy

Solar power

Convert sunlight directly to electricity. Although sunlight is free and abundant, solar electricity is still usually more expensive due to the cost of the panels. About 30% conversion efficiency are now commercially available. Over 40% efficiency has been demonstrated in experimental systems only. India has abundant sunlight. But we are far back in using sunlight to produce electricity. Land acquisition is a challenge to solar farm projects in India.

Wind power

It converts wind energy into electricity. In the Wind generation, India is fifth largest. Gujarat is leading state in wind power. Most wind turbines generate electricity from naturally occurring wind. Solar updraft towers use wind that is artificially produced inside the chimney by heating with sunlight.

Biogas

Biogas typically refers to a gas produced by the biological breakdown of organic matter anaerobically. Organic waste such as dead plants, animal bodies and byproducts and kitchen waste can be converted into a fuel called biogas.

Geothermal energy

Electricity is produced from heat coming from earth core like steam or hot water. India's geothermal energy installed capacity is experimental. Commercial use is insignificant.

Tidal wave energy

Tidal energy technologies harvest energy from the seas. The potential of tidal wave energy becomes higher in certain regions. India is surrounded by sea on three sides; its potential to harness tidal energy is significant. Tidal energy is more predicted than solar and wind energy.

Reciprocal Engines

Small electricity generators are powered by diesel, biogas or natural gas. They can provide electricity at low voltages. They are used as emergency back up for a specific facility like hospital, shopping malls, hotels, etc. to provide power during certain circumstances.

ELECTRICITY TRANSMISSION AND DISTRIBUTION

Electric power transmission is the process in the transfer of electrical power to consumers from one location to another. Transfer of electrical power from Power Stations to the industrial, commercial or residential area is necessary without interruption of flow to maintain normal lifestyle. To understand electricity transmission and distribution, we can imagine the system of blood flow in our body by arteries.

A power transmission system is sometimes referred to as a "grid", which is a network of transmission lines. The Regional Power Grids are established for getting power from different power stations. Transmission of electricity normally takes place at high voltage, i.e. more than 110 kV. Electric power is usually sent over long distances through overhead power transmission lines. Power is transmitted underground in highly populated areas, such as large cities, but is not advisable much because of much of loss of electricity in the pathway. Apart from these technical factors, thefts of electricity in pathway are also not uncommon.

The grid consists of two infrastructures: The high-voltage transmission systems, which carries electricity from the power plants and transmit it hundreds of miles away, and the low-voltage distribution systems, which draw electricity from the transmission lines and distribute it to individual customers. For transmission, high voltage is useful to minimize power loss but for distribution, high voltage lines can be proved dangerous. Electricity distribution is the second last process in the delivery of electric power, i.e. the part between transmission and user purchase from an electricity retailer. It is generally considered to include medium-voltage, i.e. less than 50kV power lines, low-voltage electrical substations and pole-mounted transformers, low-voltage distribution wiring and sometimes electricity meters. In this network, transformers are located to step down voltage before reaching to consumer.

MAINS SUPPLY

Mains is the general-purpose alternating-current (AC) electric power supply. It may be referred to as household power, household electricity, power line, domestic power, wall power, line power, AC power, city power, street power, and grid power, etc. Many different mains power systems are primarily characterized by their—
- Voltage
- Frequency
- Plugs and sockets (receptacles or outlets)
- Earthing system (grounding)
- Protection against over current damage (e.g. due to short circuit), electric shock, and fire hazards
- Parameter tolerances.

All these parameters vary among regions. The voltages are generally in the range 100–240 V (always expressed as root-mean-square voltage). The two commonly used frequencies are 50 Hz and 60 Hz.

In most countries, household power is single-phase electric power, with two or three wired contacts at each outlet.
- The live wire that is actual active contact that carries alternating current between source and machine.
- The neutral wire completes the electrical circuit by also carrying alternating current.
- The earth wire or ground connects coverings of equipment to earth ground as a protection against faults and to prevent earth shock.

Various earthing systems are used to ensure that the ground and neutral wires have the correct voltages, to prevent shocks when touching grounded objects.

Small portable electrical equipment is connected to the power supply through flexible cables terminated in a plug, which is then inserted into a fixed receptacle (socket). Giant electrical machines like fans, tube lights or industrial machine may be permanently wired to the fixed wiring so directly they can be switch on or off by switch without using plug and socket. To avoid damage to equipments because of higher voltage coming from mains due to voltage fluctuation, voltage regulator are used.

Electricity Types, Generation, Transmission, Distribution and Usage

EARTHING (FIG. 4.4)

The earth is very good conductor of electricity. In case of any fault in circuit, the additional current will flow to 'earth' through the live conductor, provided it is earthed. This is to prevent a potentially live conductor from rising above the safe level. All exposed metal parts of electrical instruments must be earthed.

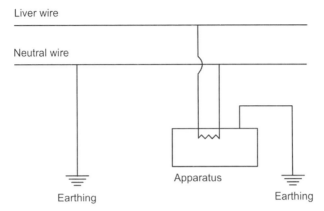

Fig. 4.4: Earthing system

As per shown in diagram, electric current is supplied from mains to load by live wire and from load to mains by neutral wire. So earthing is done from apparatus casing as well as from neutral wire to avoid earth shock.

The main purposes of the earthing are to:
- Provide an alternative path for the fault current to flow so that it will not harm the user
- Ensure that all exposed conductive parts do not reach a dangerous potential
- Maintain the voltage at any part of an electrical system at a known value so as to prevent over current or excessive voltage on the appliances or equipment.

The qualities of a good earthing system are:
- Must be of low electrical resistance
- Must be able to dissipate high fault current repeatedly.

MULTIPLE CHOICE QUESTIONS

1. In India, electricity comes in form of…
   a) Alternating current with frequency of 50 Hz
   b) Alternating current with frequency of 100 Hz
   c) Direct current with voltage 240 V
   d) Direct current with frequency of 50 Hz
2. Triboelectric effect is example of
   a) Static electricity
   b) Piezoelectric effect

c) Thermoelectric effect
d) Electrochemical effect
3. Electricity was invented by
   a) Maxwell
   b) Hertz
   c) Michel Faraday
   d) None of the above
4. Following is not renewable source
   a) Water
   b) Sun
   c) Wind
   d) Petrol
5. Disadvantage of solar energy is
   a) It is free
   b) It is abundant
   c) It is converted to electrical energy only upto 30%
   d) All of the above
6. Concept of central power station is established in
   a) 1820
   b) 1881
   c) 2006
   d) 2009
7. Reciprocal engines are useful mainly for
   a) Running a vehicle
   b) Back up for hospital
   c) Running a industry
   d) All of the above
8. Following are types of electricity
   a) Static electricity
   b) Current electricity
   c) a and b
   d) Alternating current and direct current
9. When balloon is rubbed with wool, balloon will of which polarity?
   a) Negative
   b) Positive
   c) Cannot be commented
   d) Neutral

Answers:

1—a	4—d	7—b
2—a	5—c	8—c
3—c	6—b	9—a

Bibliography:
1. Robinson J, Mackler S. Clinical Electrophysiology Electrotherapy and Electrophysiological Testing, 2nd ed. Williams and Wilkins, 1989.
2. Clayton's electrotherapy. AITBS Publishers, 2002.

Reference websites:
1. http://en.wikipedia.org
2. http://www.amasci.com
3. http://www.diyfixit.com
4. http://www.earthinginstitute.net
5. http://www.electricityforum.com
6. http://www.mapsofindia.com
7. http://www.ofgem.gov.uk
8. http://www.power.indiabizclub.com

Sources of Direct Current | 5

CELL

Primary Cell or Electric Cell or Electrochemical Cells or Galvanic Cell or Voltaic Cell or Disposable Cell

Most commonly word cell is used for battery which we are using in routinely in torch, remote controller of television, etc. It is the electric device which converts chemical energy into electrical energy irreversibly. This cannot be used again. First electric cell was made in 1799 by the Italian Scientist Volta.

They are primarily made up of copper and zinc plates with liquid of sulfhuric acid. They also have two poles named cathode (negative polarity) and anode (positive polarity). When the electric cell becomes a part of external circuit, they give small amount of direct current.

Secondary Cell or Storage Cell or Rechargeable Cells

These cells are designed in such a way they can be used again. They need to get charged once they are used. In charging, chemical process which occurs during discharging of cell is reversed and cell is ready to be used again.

Photovoltaic Cells or Photoelectric Cells or Solar Cells

Cells are available in large variety. They convert light energy into electrical energy.

Fuel Cells

Fuel cells will perform chemical reaction in fuel and produces electrical energy. Mostly they are used in vehicles.

Electrolytic Cells

They break the chemicals used such as water molecule can be split into hydrogen and oxygen to use as fuel.

Hazards

Blast

If attempt is made to charge primary cell or overcharging of secondary battery is done, explosion of battery may occur.

Leakage

If leakage of chemicals from cell occurs, it may be proved poisonous.

Environmental concern

Used battery may be increase electronic waste and chemical used may spoil earth and air by spreading pollution.

Children concern

Small button cells can be swallowed by young children and also by some animals if they are thrown in other waste materials.

THERMIONIC VALVES

They are electronic devices consisting of a system of electrodes arranged in an evacuated glass or metal envelopes and used for rectification of alternating current.

DIODE

It is electrical device which will allow the flow of electron only in one direction. They are working as one-way valves and used in various circuits. There is exception of zener diode which will also allow the current to flow in reverse direction when voltage reaches a particular threshold. They are made up from silicon, germanium, or selenium. They have cathode and anode.

When dopants are added to the any side of a semiconductor, one side becomes n-type semiconductor (negative) and other side becomes p-type semiconductor (positive) and p-n junction is formed. This is known as junction diode or a semiconductor diode. The most common type of diode is silicon diode (Fig. 5.1).

- **Forward and reverse biasing:** When any source of electric energy is connected with diode, it is known as biased.
- **Bias voltage:** In above mentioned example, applied voltage is known as bias voltage.

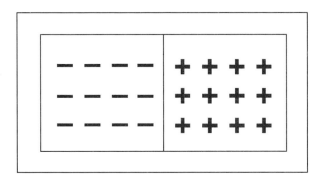

Fig. 5.1: Junction diode

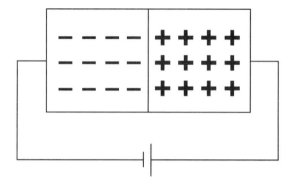

Fig. 5.2: Forward bias

- **Forward bias**: When negative terminal of battery is connected with n-type and positive terminal of battery is connected with p-type, it is said to be forward biased (Fig. 5.2).

  The external voltage overcomes the junction potential and provides an easy pathway for the flow of charge carriers across the junction. The flow of electrons occurs from n to p and flow of positive holes occurs from p to n. The device offers low resistance.

- **Reverse bias**: When negative terminal of battery is connected with p-type and positive terminal of battery is connected with n-type, it is said to be reverse biased (Fig. 5.3).

  In this case the charge carriers are repelled away from the junction and no current flows through the junction. The device offers high resistance. There will be a small current due to few charge carriers, e.g. electrons in p type and holes in n type regions.

- **Uses**:
  - Rectifiers
  - Signal limiters
  - Voltage regulators
  - Switches, etc.

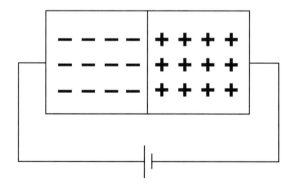

Fig. 5.3: Reverse bias

- **Symbol**
  - ▬⋯⋯▷⊢⋯⋯

TRIODE

It is electron tube made up of three electrodes mounted in metal or glass container as follows:
- Cathode as filament
- Anode as plate
- Control grid.

Uses

In electronic circuit, they are used as an amplifier for audio and radio signals. Amplification is done by grid so it is essential part of triode.

Variants

Variants of triodes are tetrode or pentode, etc. with extra number of grids.

RECTIFIER

It is an electrical device that converts alternating current (AC) into direct current (DC). This process is known as rectification. The electrical device that converts direct current (DC) into alternating current (AC) is known as an inverter.

Half-wave Rectification (Fig. 5.4)

In half-wave rectification one phase of alternating current, either the positive or negative phase of the AC wave is allowed to pass and another half is not allowed to pass. Only half of inputs of alternating current reaches for output so naturally mean output voltage decreases. In half-wave rectification single diode in a single-phase supply and three in a three-phase supply are needed. As only one side of the phases either positive or negative phases are gained in output so current will be pulsed direct current. Much more filtering is needed to eliminate harmonics of the AC frequency from the output.

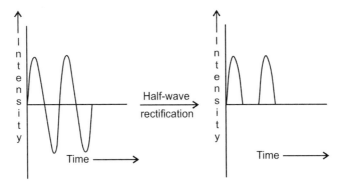

Fig. 5.4: Half-wave rectification

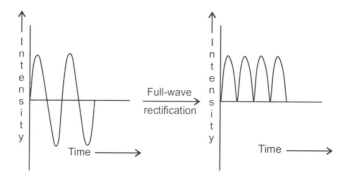

Fig. 5.5: Full-wave rectification

Full-wave Rectification (Fig. 5.5)

In full-wave rectification conversion of the whole of the input waveform of alternating current to direct current occurs. It means it converts both polarities of the input waveform, i.e. positive and negative phases of alternating current to direct current. As in full-wave rectification, all positive and negative phases of inputs alternating current reaches for output so naturally mean output voltage will be more as compared to half-wave rectification. Two diodes and a transformer, or four diodes are needed.

Uses

- To get the direct current from alternative current
- They are component of power transmission systems
- They can also be used for detection of radio signals.

CHOKE COIL (FIG. 5.6)

A circuit element used to suppress or limit the flow of alternating current without affecting the flow of direct current. It is a coil of wire of insulating material, wound on a magnetic core. It is used as a passive inductor which blocks higher-frequency alternating current in an electrical circuit while passing signals of much lower frequency alternating current and direct current.

Fig. 5.6: Choke coil

Sources of Direct Current

The name comes from choking means blocking high frequencies while passing low frequencies. Choke coils are used in a number of electrical devices. During the process of sending a signal through a circuit, it is necessary to allow the wanted signals to pass and block the unwanted signals. Without a choke coil, the power line creates a lot of electrical noise as it traveled to its destination.

The choke is an inductor. It works on the principle of self-induction. When the current passing changes, as AC currents do, it creates a magnetic field in the coil which works against that current producing it. This property blocks most of the AC from passing through the circuit. So currents which do not change, such as DC currents can continue. Its forward EMF and backward EMF elevates through and suppress crests of variable intensity direct current respectively so it is used to smoothen of direct current.

MULTIPLE CHOICE QUESTIONS

1. Pick the odd out
    a) Electric cell
    b) Electrochemical cells
    c) Galvanic cell
    d) Storage cell
2. Primary cell when connected to circuit it gives...
    a) Direct current
    b) Alternative current
    c) Both of the above
    d) None of the above
3. Light energy is converted into electrical energy by:
    a) Fuel cells
    b) Electrolytic cells
    c) Solar cells
    d) Storage cells
4. Diodes are made up of...
    a) Silicon
    b) Germanium
    c) Selenium
    d) All of the above
5. When negative terminal of battery is connected with n-type and positive terminal of battery is connected with p-type, it is said to be in:
    a) Forward biased
    b) Reversed biased
    c) Junction biased
    d) Nil biased
6. The device that converts direct current into alternating current is known as:
    a) Rectifier
    b) Inverter

c) Rheostat
d) Transistor
7. Mean output will be more in which of the following?
   a) Half-wave rectification
   b) Full-wave rectification
   c) Equal in both of the above
   d) Depends on voltage of load
8. The function of choke coil is…
   a) To block high frequency current
   b) To block low frequency current
   c) To block direct current
   d) To block all types of current
9. Pick up correct statement for triode
   a) Cathode as plate, Anode as filament, Control grid
   b) Cathode and Anode as plates, Control grid
   c) Cathode as filament, Anode as plate, Control grid
   d) Cathode Anode as filaments, Control grid

Answers:

1—d	4—d	7—b
2—a	5—a	8—a
3—c	6—b	9—c

Bibliography:
1. Mohan S, Smitha KG. MCQs in Electrotherapy, 1st ed. Jaypee Brothers Medical Publishers (P) Ltd, 2006.

Reference websites:
1. http://en.wikipedia.org
2. http://simplewayoflearnphysics.blogspot.in
3. http://www.blurtit.com
4. http://www.cjseymour.plus.com
5. http://www.rapidtables.com
6. http://www.standrews.ac.uk
7. http://www.thefreedictionary.com
8. http://www.wisegeek.com

Skin Resistance, Electrodes and Gel 6

SKIN RESISTANCE

Our skin is made up of different layers, e.g. epidermis and dermis. Epidermis has again different five layers. These skin layers are offering resistance to flow of current from electrode to patient's body.

Measurement of Skin Resistance

'Galvanic skin response', also known as 'psychogalvanic reflex' is the method to measure skin conductance for electric current. Conductance is inversely proportional to resistance. So more the conductance, less the resistance and less the conductance, more the resistance.

Factors Affecting Skin Resistance

Cleanliness of skin

The resistance of human skin is highly variable depending on several different variables, but the two main variables are whether the skin is clean or dirty. Clean skin often has a resistance of about 500 ohms. Dirty skin can have electrical resistances of upto several million ohms.

Wetness of skin

If skin is wet with water or anything, skin resistance will be lower. For this reason only with wet hand we should not operate switch of any electric machine as chances of electric shock are more with wet hands.

Autonomic nervous system

It varies with emotion, attention, and stress. The sympathetic nervous system is subconsciously activated when the body is subjected to any stress. Autonomic nervous system activity causes a change in the skin's conductivity. Higher arousal will almost instantaneously cause a fall in skin resistance; reduced arousal will cause a rise in skin resistance. When patient is under mental stress sweat glands become more active and secretion of sweat will increase. This increased secretion of sweat will make skin more wet leading to reduction of skin resistance. When skin is dry it will offer more resistance.

Capacitance and frequency of electric current

Formula for skin resistance is
- $Z = 1 / 2\pi fC$

- Where π is constant of approx. value 3.14
- f is frequency of current (in hertz)
- C is capacitance (in faraday).

Skin resistance is inversely proportional to capacitance and frequency. If capacitance will increase, skin resistance will decrease and if capacitance will decrease, skin resistance will increase. Similar will be applicable for frequency. For high frequency current like short wave diathermy skin resistance will be very less so diathermy waves will pass throughout. But for transcutaneous electrical Nerve Stimulation frequency will be low so it cannot penetrate more in the body.

Clinical Application

In electrotherapy, before treating patients we should clean the skin with alcohol to remove dust particles, lotion, cream applied over it. These things may offer more resistance to electric current. Remember that natural skin resistance offered by structure of skin cannot be eliminated but only resistance offered by externally applied agents can be minimized. And while giving electrical stimulation electrode should be dipped in water to minimize skin resistance by making it wet.

ELECTRODES

Definition

An electrode is an object made up of conductive material which works as interface between patients' body and electric stimulator machine.

Connection of Electrode with Machine

Connection of electrode and machine is made with wire with insulated coverings. They are known as leads, cables or cords.

Classification of Electrode According to its Placement Site in Body

- Surface electrodes: Electrodes are attached to skin
- Invasive or indwelling electrodes: Electrodes are implanted near the nerves or bones
- Internal electrodes: Electrodes are inside the body cavities.

Physiotherapists are using surface electrodes only so our discussion will be for surface electrodes only in detail.

Types of Surface Electrodes

In different clinical settings different types of electrodes are used depending on individual preference. Each type has its merits and demerits.

Rubber electrodes (Fig. 6.1)

They are made up of carbon and silicone. These electrodes are used most commonly. Their color is mostly black. They have hole to connect it with

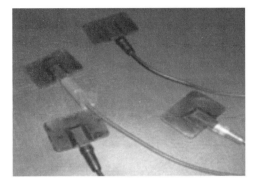

Fig. 6.1: Rubber electrodes

lead on one side. By shape they are rectangular, square or circular depending on manufacturer.

Advantages
- They are user-friendly.
- They can be adjusted according to body contour making firm contact.

Disadvantages
- They need gel or water to make contact between skin and electrode. Without this conducting medium current will not enter to patient's skin.
- They become very dirty and less flexible to make firm contact with patients' body with more use. They must be disinfected with sterilizing material after every use.
- They may crack in between after sometime.
- They need stabilization with strap of adhesive tape on the skin.

Metal electrode (Fig. 6.2)
- They are made up of metals as the name suggests. They are used less frequently.

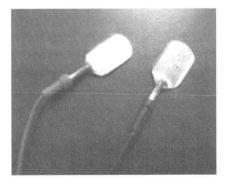

Fig. 6.2: Metal electrodes

Advantages
- They are also easy to use
- They are not breakable
- They do not need conducting gel.

Disadvantages
- They need wet lint pad made up of cloth that surrounds electrode on each side as direct contact of electrode with skin leads to skin chemical burn.
- They are not flexible to body contour.
- They also need stabilization with strap or adhesive tape.

Vacuum electrodes

They are cup-like structure made up of plastic. Inside this cup-like structure, there will be electrode. Vacuum will be created by machine between cup and body so electrode will be held over the body part.

Advantages
- They do not need stabilization by strap of tap
- It can be placed over the body's irregular body part
- They also do not need conducting gel.

Disadvantage
- Cost of machine will be higher because of extra vacuum component.

Self-adhesive electrodes

They are adhered to patients' body by sticky material on it. These are most recent advance used nowadays.

Advantages
- They are very easy to use
- They do not need any type of stabilization material.

Disadvantage
- They tend to wear after repeated use.

Parameters of Electrodes

Size

It depends on area to be treated. For larger area to stimulate, electrode size should be large. For example, to stimulate quadriceps muscle large electrode is required as small electrode will stimulate only few numbers of muscle fibers in these muscles. For smaller area of treatment, small electrode should be used. For example, to stimulate one muscle on hand-like abductor pollicis brevis. If large electrode is used for this muscle, it will lead to other surrounding undesired muscle contraction rather than that of desired muscle.

Size also depends on the technique. Like if unipolar technique is used in which one electrode is active and other electrode is indifferent, size of active electrode is lesser so give more current density (intensity of current per unit area) and size of indifferent electrode is larger to reduce current density on area where contraction is not desired but just to serve the purpose of completing circuit. For bipolar technique, both electrodes are having equal importance and they are placed on target area. For this case, size of body electrodes will be same.

Shape

Shape of electrode depends on manufacturing company. Most commonly available shapes are circular and rectangular with rounded edges. These designs are believed to be safe as there is no concentration of current at sharp angles of electrodes.

Location

Placement of electrode should be explained with perfect landmark. For example, to stimulate tibialis anterior muscle, electrode is placed over the junction of upper one-third and lower two-thirds of leg lateral to anterior border of tibia, i.e. shin. If distance is fixed like 5cm above the upper lateral border of patella; distance should be measured from center of electrode to upper lateral border of patella.

If unipolar (or monopolar) technique is used, one electrode will be places over the target area which is to be stimulated. This electrode is known as stimulating or active electrode. And another electrode is placed over convenient are of subject's body to complete circuit. This second electrode is known as indifferent (or dispersive or reference) electrode. Some practitioners are using word ground electrode for this electrode. This is wrong terminology as it is not connected with ground.

ELECTRODE GEL

A gel is a solid, jelly-like material that can have properties ranging from soft and weak to hard and tough. Its meaning is cold and immobile. On contact it is felt cool and by property it is immobile when put on surface unlike liquid. Gel is known as coupling medium of ultrasound and electric current with patient's body surface. Applying a gel to the skin during an ultrasound or using carbon electrodes in electrotherapy allows the ultrasound and electrical current to pass more easily from ultrasound transducer or carbon electrode to the patient's skin. Using a gel during treatment will allow ultrasound and current to pass without much interference. Air does not transmit ultrasound so ultrasound will be reflected from air. Because of this reason ultrasound is not useful for diagnosis of lung conditions as air will be present always in lungs. While giving ultrasound if gel is not used, air between transducer head of ultrasound and skin will reflect the ultrasound to transducer head again and may damage it also.

Characteristics

- Air bubbles should not be present in gel because air is reflector for ultrasound and insulator for current so presence of air will stop the transmission of ultrasound and current between electrode and body.
- Gel should be clinically tested not to produce any skin reactions.
- Ultrasound gel is typically clean and thick, but should not be excessive sticky.
- Some patients feel this gel very cold so they might not be tolerating. So some professionals use ultrasound gel warmers to reduce discomfort for patients and create a more pleasant ultrasound experience.

MULTIPLE CHOICE QUESTIONS

1. Following is not important for selection of electrodes
   a) Size
   b) Shape
   c) Color
   d) All of the above
2. Following electrodes cannot be adjusted according to contour of body part of subject
   a) Surface electrodes
   b) Carbon electrodes
   c) Vacuum electrodes
   d) None of the above
3. Identify the correct statement
   a) For unipolar technique both electrodes are having same size
   b) For unipolar technique both electrodes are having same shape
   c) For bipolar technique current density will be unequal on placement of both electrodes
   d) None of the above
4. Use of following electrodes will not require any type of stabilization of electrodes with strap of adhesive tape
   a) Surface electrodes
   b) Carbon electrodes
   c) Vacuum electrodes
   d) All of the above
5. Use of electrode gel is as
   a) Topical medication
   b) Conducting medium
   c) Both of the above
   d) None of the above
6. Identify incorrect
   a) Electrode gel should not be having air bubbles
   b) Electrode gel should not be very sticky
   c) Electrode gel should not be very cold
   d) Electrode gel should not transmit ultrasound and current

7. Skin resistance will be the least for following current
   a) High frequency current
   b) Medium frequency current
   c) Low frequency current
   d) Skin resistance has no relation with frequency of current
8. To minimize the skin resistance
   a) Skin should be shaved off for hair
   b) Skin should be cleaned with alcohol
   c) Skin should be made wet
   d) All of the above

Answers:

1—c	4—c	7—a
2—a	5—b	8—d
3—d	6—d	

Reference websites:
1. http://bio-medical.com
2. http://uk.answers.yahoo.com
3. http://wiki.answers.com
4. http://www.ehow.com
5. http://www.wisegeek.com

Electric Shock 7

The human body is made up of 60–70% of water. This makes it a good conductor of electricity. Electric shock is painful stimulation of sensory nerves and sometimes also motor nerves. That occurs upon contact of a body with any source of electricity that causes a sufficient current through the skin, muscles, or hair. Typically, the expression is used to denote an unwanted exposure to electricity, hence the effects are considered undesirable.

Shock is a common occupational hazard associated with working with electricity. Physiotherapy department is full of electronic equipments so if electric shock may happen at any moment, steps can be taken immediately. Electric shock will be affected by type of electric current that is explained in Table 7.1.

Table 7.1: Difference between electric shock caused by direct current (DC) versus alternating current (AC)

Direct current (DC)	Alternating current (AC)
A person can feel at least 5 mA for DC	A person can feel at least 1 mA of AC at 60 Hz
The current may cause tissue damage or fibrillation which leads to cardiac arrest at 300–500 mA of DC	The current may cause tissue damage or fibrillation which leads to cardiac arrest at 60 mA of AC
Severity of electric shock will be less severe as intensity of current is constant	Severity of electric shock will be more severe as intensity of current is continuously changing

Causes of Electric Shock

- Sudden onset flow of current through body
- Sudden termination of current already passing through body
- Sudden large variation in current already passing through body.

Types of Electric Shock According to Point of Entry

Macroshock

Macroshock occurs when current passes between two contact points on the skin. If the voltage is less than 200 V, then the human skin provides resistance to the electric current. If the voltage is above 450–600 V, then breakdown of the skin occurs. The protection offered by the skin is lowered by sweating also because wet part will reduce the skin resistance.

Microshock

If person gets electric shock when an electrical circuit is established by electrodes introduced in the body, bypassing the skin is known as microshock. In this case, potential for lethality is much higher if a circuit through the heart is established. This is known as a microshock. Currents of only 10 µA can be sufficient to cause fibrillation in this case. This happens when the patient is connected to multiple devices, like cardiac catheters, pacemaker, etc. in hospital settings.

Severity of Shock

- In mild electric shock victim has no any type damage to his body's structure or function.
- In moderate type, victim may develop hypotension because of vasodilatation after electric shock and also unconsciousness.
- In severe cases, there may be termination of respiration and heart function.

Signs and Symptoms

Burns

Heating due to resistance can cause extensive and deep burns. Voltage levels of 500–1000 volts tend to cause internal burns due to the large energy. Damage due to current is through tissue heating.

Ventricular fibrillation

Ventricular fibrillation or cardiac arrest can occur at different intensities depending on type of current as explained above. Electric shock can cause fibrillation. If not immediately treated by defibrillators, it may prove lethal. Death caused by an electric shock is called electrocution.

Neurological effects

Current can cause interference with nervous control, especially over the heart and lungs. Repeated or severe electric shock which does not lead to death has been shown to cause neuropathy. Numbness or tingling and change in vision, speech, or in any alteration in sensations may occur.

Musculoskeletal effects

Muscle spasms or contractions, sudden immobility or fractures or a body part may look deformed.

More

Other fatal complications include interrupted breathing, irregular heartbeats, chest pain, seizures and unconsciousness.

Body Resistance

The voltage necessary for electrocution depends on the current through the body and the duration of the current. Ohm's law states that the current drawn depends on the resistance of the body. The resistance of human skin varies from person to person and fluctuates between different times of day. The resistance offered by the human body may be as high as 100,000 ohms. Wet or broken skin may drop the body's resistance to 1,000 ohms.

Factors in Lethality of Electric Shock

The lethality of an electric shock is dependent on several variables:

Intensity of current

The higher the current, the more likely it is lethal. Since current is proportional to voltage when resistance is fixed (Ohm's law), high voltage is an indirect risk for producing higher currents.

Duration

The longer the duration, the more likely it is lethal. Safety switches may limit time of current flow.

Pathway

If current flows through the heart muscle, it is more likely to be lethal.

Very high voltage

This is an additional risk over the simple ability of high voltage to cause high current at a fixed resistance. Very high voltage, enough to cause burns, will cause dielectric breakdown at the skin, actually lowering total body resistance and, ultimately, causing even higher current than when the voltage was first applied. Contact with voltages over 600 volts can cause enough skin burning to decrease the total resistance of a path though the body to 500 ohms or less.

Resistance of skin

Less the resistance, more the severity of electric shock. Wet body part will have less resistance so more chances of electric shock.

Type of current

Shock is more severe with alternating current as intensity of current is not constant so there is no accommodation by patient which leads to strong sensory stimulation.

Frequency

It may cause cardiac arrest or muscular spasms. Very high frequency electric current causes tissue burning, but does not penetrate the body far enough to cause cardiac arrest.

Medical Uses

Electric shock is also used as a medical therapy, under carefully controlled conditions:
- Electroconvulsive therapy is a psychiatric therapy for mental illness.
- As a surgical tool for cutting or coagulation.
- As a treatment for fibrillation or irregular heart rhythms.
- As a method of pain relief, e.g. transcutaneous electrical nerve stimulator.

Torture

Electric shocks are used as a method of torture, since the received voltage and current can be controlled with precision and used to cause pain and fear without physically harming the victim's body.

Self-Care / First Aid / Treatment

- Do not touch the person until power is shut off.
- If the source is a low-voltage current, remove the fuse or switch off the circuit.
- If there is no switch in the circuit, use a board, wooden stick, rope, etc. to get the person away from the source. But do not touch the victim with your hand.
- Unless it is absolutely necessary, don't move the person. He or she could have a traumatic injury, especially to the head or neck.
- Check for burns. Cover burned areas with dry, sterile dressings.
- After minor shock, allow the victim to rest.
- Give something to drink to prevent hypovolemia because of vasodilatation blood pressure lowers. Hot fluids must not be given to drink as it will lead to further vasodilatation and further fall in blood pressure.
- Tight clothing should be loosened and plenty of air should be allowed. Fans should be switched on or windows should be opened. Excessive heat will increase metabolism in the tissue and demand of oxygen will increase.
- If victim has stopped breathing, artificial respiration either manually or with Ambu bag will be helpful. Do not give artificial respiration if patient is breathing naturally. Oxygen therapy may be helpful in some cases.
- If victim's pulses are absent, heart may not be functioning. External cardiac massage must be started without delay.
- If heart is under fibrillation, only cardiac massage will not be helpful. For this case, instrument called defibrillator should be used.
- Even if victim is awake and has no problem after electric shock, he should be moved to hospital for check-up and in major case medical officer should be called without delay.

Prevention

- All apparatus should be checked before use.
- All connections should be checked before use.

- Controls should be at zero before starting apparatus.
- Proper warm up time should be given to apparatus before starting treatment.
- Intensity should be increased gradually.
- Patient is not allowed to touch circuit during treatment.
- All apparatus should be serviced at regular interval by qualified engineer.
- Covering of apparatus should be of insulating material.
- Stay clear of fallen wires. Inform the police, electric company, etc.
- Don't turn electrical switches on or off or touch an electric appliance while your hands are wet, while standing in water, or when sitting in a bathtub.
- Replace worn cords and wiring.
- Cover all electric sockets with plastic safety caps.
- Before you do electrical repairs, remove the fuse from the fuse box or switch off the circuit breaker.

EARTH SHOCK

When a person comes in contact with live wire and earth together, he may receive an electric shock that is known as earth shock.

Circuit

Earth is very good conductor of electricity. From live wire current enters into victim's body and from there it will enter into earth. So victim become a part of circuit and receive an earth shock.

Connection to Live Wire can be Made in Following Conditions

- By touching part of circuit when current is passing through it.
- If covering of lead with insulating material is not proper and conducting part is exposed to environment so patient may touch this exposed part of wire.
- If wire comes in contact with metal casing of any instrument.
- If switch does not break live wire, current may pass even when switch is off.

Connection to Earth can be Made in Following Conditions

- By touching conductor which is contact with earth, e.g. gas lines or water pipes, etc.
- By putting patient on metal bed which is on the earth.
- Especially when floor is damp.

Examples of Earth Shock

If connection to live wire and earth is made together in above mentioned ways, victim may get earth shock.

Precautions Against Earth Shock

- Department should be designed in such a way with earthing facility so that no chances of earth shock.
- Floor should not be wet.
- Floor should be of insulating material. If not so carpet should be spread to avoid direct contact with earth.
- Even in floor made up of insulating material, care should be taken that water does not leak through cracks in tiles to make earth connection.
- Physiotherapist and patient should wear footwear like shoes, slippers, etc. to avoid direct connection of body with earth.
- Gas lines and water pipes should not be inside the department. These lines should be connected from exterior of department.
- Switches must be there in each electric line for each modality and they must break live wire.
- Fuses must be implanted in department.
- When treatment is given in water bath, container for water should be of insulating material if it is placed on earth or any conducting surface.
- Container of water should not be leaking. Water should not be added when current is flowing.
- To make current earth free, either current from batteries should be used or alternating current can be rendered by a static transformer.
- Patient is not supposed to touch any part of machine during the treatment.

MULTIPLE CHOICE QUESTIONS

1. Electric shock is severe when it passes though
   a) Head, neck and heart
   b) Lower limbs
   c) Lower trunk and limbs
   d) It is irrespective of pathway taken by current
2. An electric shock is more severe when
   a) The intensity of current through the body is less and the resistance of the body is more
   b) The intensity of current and resistance offered is less
   c) Intensity of current is more and resistance is less
   d) Severity does not depend on current or the resistance of the body
3. When oral fluid is offered to victim of electric shock, care should be taken to see that
   a) Only water is given
   b) Only hot fluid is given
   c) No hot fluid is given
   d) Both cold and hot fluid can be given
4. A person receives earth shock when
   a) He touches the exposed part of the apparatus casing
   b) He touches the water pipe nearby

c) When he brings both active and inactive electrodes touching each other
d) All of the above

5. An important precaution against earth shock is
   a) Patient should not be in contact with earth and live wire simultaneously
   b) Footwear should be worn
   c) Use of 3 pin power plugs
   d) All of the above

6. In the shock, immediately you should
   a) Put victim down and allow to rest
   b) Give some hot fluids to drink
   c) Disconnect the victim from source of supply
   d) Immediately pull victim by hand and bring him to safety

7. The most severe effects of electric shock
   a) Fall in BP, faintness and anxiety
   b) Fall in BP, headache and stress
   c) Fall in BP, unconsciousness, respiratory and brain death
   d) Fall in BP, unconsciousness, respiratory and cardiac arrest

8. In a three-pin plug, the internationally accepted system of coding is
   a) Red to live, black to neutral and green to earth
   b) Green to live, brown to neutral and black to earth
   c) Brown or black to live, blue or red to neutral and green or yellow to earth
   d) Green to live, red to neutral and blue or yellow to earth

9. Sudden cessation of current can also cause a shock when the current used is…
   a) Direct current
   b) Alternative current
   c) Both of the above
   d) It is irrespective of type of current

10. Giving hot fluid to a shock victim causes
    a) Vasodilatation and further fall in BP
    b) Vasoconstriction and increase in BP
    c) Vasodilatation and increase in BP
    d) Vasoconstriction and fall in BP

11. A direct connection between earth and live wire of main results in…
    a) Electric shock
    b) Earth shock
    c) Short circuit
    d) Sparking

Answers:

1—a	5—d	9—a
2—c	6—c	10—a
3—c	7—d	11—b
4—d	8—c	

Bibligraphy:
1. Clayton's Electrotherapy. AITBS Publishers; 2002.
2. Mohan SS, Smitha KG. MCQs in Electrotherapy, 1st ed. Jaypee Brothers Medical Publishers (P) Ltd.

Reference websites:
1. http://en.wikipedia.org

Magnet and its Properties   8

INTRODUCTION

A magnet is an object that produces a magnetic field which has properties of power of attraction and repulsion for certain materials and tendency when free to rotate to come to rest position in north south direction only.

Nature of magnet is such that in magnetic materials, all the molecules are working as magnet. Moreover, they are well-arranged as shown in Figure 8.1b. If we break a magnet into thousands of small pieces, each piece including the smallest piece will work as separate magnet. In all materials other than magnet, this arrangement is not possible as all molecules are arranged randomly as shown in Figure 8.1a. But by certain techniques these randomly arranged molecules can be arranged in well-organized manner and magnet can be produced (Figs 8.1a and 8.1b).

The region around a magnet where the force of attraction or repulsion can be detected is called magnetic field. Magnetic field around a magnet can be detected by using a magnetic compass.

Fig. 8.1a: Unmagnetized material   **Fig. 8.1b:** Magnetized material

GEOMETRY OF MAGNET

Geometric Length

In Figure 8.2 actual length of bar magnet AB is known as geometric length of magnet.

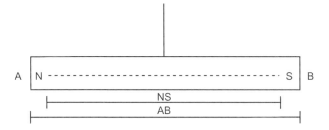

Fig. 8.2: Geometry of magnet

Magnetic Length

Actual magnetic poles are not situated at two geometric ends. They are situated slightly inside of two ends. This distance NS is known as magnetic length. Magnetic line occupies only 6/7th of geometric length.

Equatorial Line

The line passing through the center of magnet and perpendicular to long axis of magnet is known as equatorial line.

Magnetic Meridian

A vertical plane passing perpendicular to long axis of magnet is known as magnetic meridian.

TERMINOLOGIES

- Ferromagnetism is the phenomenon exhibited by materials like iron, nickel or cobalt that become magnetized in magnetic field and retain their magnetism even after removal of magnetic field.
- Diamagnetism is the phenomenon exhibited by materials like copper or bismuth that become magnetized in a magnetic field with polarity opposite to the magnetic force.
- Paramagnetism is the phenomenon exhibited by aluminum or platinum those become magnetized in a magnetic field but magnetism disappears when the field is removed.

TYPES

Based on its manufacture mainly magnets are of two types.
- Natural magnets
- Artificial magnets.

Comparison between natural magnets and artificial magnets is given in Table 8.1.

Table 8.1: Comparison between natural magnets and artificial magnets

Natural magnets	Artificial magnets
They are found from earth	They are made with manual procedure

Contd...

Contd...

Natural magnets	Artificial magnets
Their strength of magnetism is weaker	Their strength of magnetism is stronger
They are made up of iron, steel, cobalt, nickel, etc.	They are made from iron or steel
They are of irregular shape and size.	They can be made of any size and shape. For example bar magnet, horse-shoe magnet, magnetic needle, compass needle, etc.

OTHER TYPES OF MAGNETS

- Permanent magnets
- Temporary magnets
- Electromagnets.

Permanent Magnets

They are retaining their certain degree of magnetism forever after they are magnetized. Mostly we use these types of magnets at household doors and also refrigerator doors. These types of magnets will maintain position of doors in their place. Different types of permanent magnets differ in characteristics about demagnetization, strength and its variation with temperature. They are generally made up of ferromagnetic materials.

Temporary Magnets

Temporary magnets are having properties of magnetism for some period of time when they are under influence of strong magnetic field. They will lose their magnetism when this strong magnetic field disappears. They act like permanent magnets only when they behave as magnets. Examples would be paperclips and nails and other soft iron items. They are used in telephones, electronic motors, etc.

Electromagnets

Electromagnets are produced when electric current produce magnetism in certain material with formation of magnetic field around that. They will be working as electromagnets only when current passes in near space. When current stops to flow they will lose their magnetic properties. They are extremely strong magnets. They are produced by placing a metal like iron inside a coil of wire carrying an electric current. This current produces magnetism in the iron. The strength of the magnet is directly proportional to the intensity of the current and the number wires in the coil. Its polarity depends on the direction of flow of current. They are used in large cranes for lifting heavy iron cables and rods for construction. They are also used in many electronic instruments like computers, television,

Magnet and its Properties 59

telephones, motors, generators, loudspeakers, hard disks, MRI machines and scientific instruments, etc.

SHAPES

Permanent magnets can be made into following shapes:
- Round bars
- Rectangular bars
- Horseshoes
- Rings
- Disks
- Rectangles
- Other custom shapes.

MAGNETIC EFFECT OF ELECTRIC CURRENT

As explained earlier when current passes through any conductor, it produces magnetism in surrounding object. Today we are using motor based on this principle only. A current carrying conductor creates a magnetic field around. The nature of the magnetic field lines around a straight current carrying conductor is concentric circles with center at the axis of the conductor. The strength of the magnetic field depends on the current passing through the conductor. If the conductor is in the form of a circular loop, the loop behaves like a magnet. A current carrying conductor in the form of a rectangular loop behaves like a magnet and when suspended in an external magnetic field experiences force. See electromagnets (P. 58).

ATOMIC/MOLECULAR THEORY OF MAGNETISM

The molecular/atomic theory of magnetism was given by Weber. According to this theory:
- Every molecule of a magnetic substance (whether magnetized or not) is a complete magnet in itself, having a North pole and a South pole of equal strength.
- In an unmagnetized substance, the molecular magnets are randomly oriented such that they form closed chains. The North pole of one molecular magnet cancels the effect of South Pole of the other so that the resultant magnetism of the unmagnetized specimen is zero.
- On magnetizing the substance, the molecular magnets are realigned so that North poles of all molecular magnets point in one direction and South poles of all molecular magnets point in the opposite direction.
- When all the molecular magnets are fully aligned, the substance is said to be saturated with magnetism.
- At all the stages, the strengths of the two poles developed will always be equal.

- On heating the magnetized specimen, molecular magnets acquire some kinetic energy. Some of the molecules may get back to the closed chain arrangement. That is why magnetism of the specimen would reduce on heating.

PROPERTIES OF A MAGNET

- Earth itself is big magnet. When magnet is free to move, North Pole of magnet is pointing to North Pole of earth and South Pole of magnet is pointing towards South Pole of earth.
- Like magnetic poles repels each other and unlike magnetic poles attracts each other. For example when two North or two South magnetic poles come nearer to each other, they will be repelled. When North and South magnetic poles come near to each other, they will be attracted toward each other.
- When magnet is rubbed over certain materials, it will produce magnetic properties in that material by realigning molecules in that. In some of the materials without contact also magnet can produce magnetic properties, this is known as magnetic induction.
- Magnet will attract certain materials like iron, steel, nickel, cobalt, etc.
- Magnet will produce one attractive field around it, in which if other object will come, it will feel some force. This is known as magnetic field which has certain properties.
- The force of attraction or repulsion of a magnet is greater at its poles than in the middle.
- The force of attraction or repulsion between two magnets will be directly proportional to product of two magnetic poles strength and inversely proportional to square of distance between two magnets.

PROPERTIES OF MAGNETIC LINES OF FORCE

- Magnetic lines of force start from the North Pole and end at the South Pole outside the body of magnet (Fig. 8.3).

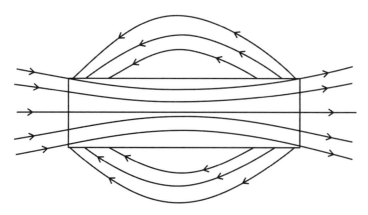

Fig. 8.3: Magnetic lines of force

Magnet and its Properties

- They are continuous through the body of magnet.
- Magnetic lines of force can pass through iron more easily than air.
- They tend to contract longitudinally.
- They tend to expand laterally.
- They never cross one another.
- They all have the same strength.
- Their density decreases (they spread out) when they move from an area of higher permeability to an area of lower permeability.
- Their density decreases with increasing distance from the poles.
- They are considered to have direction as if flowing, though no actual movement occurs.
- They seek the path of least resistance between opposite magnetic poles.
- They attempt to form closed loops from pole to pole.

Comparison between electricity and magnetism is given in Table 8.2.

Table 8.2: Comparison between electricity and magnetism

Electricity	Magnetism
Two charges exist: Positive and negative	Two poles exist: North and South
Like charges repel each other and unlike charges attract each other. Electric charge may exist separately. For example positive or negative charge	Like poles repel each other and unlike poles attract each other Magnetic poles always are in pair. Separate north or South pole does not exist
Electric lines of force start from positive charge and end into negative charge. Thus they are not in close loop	Magnetic lines of force are close continuous curve passing throughout the body of magnet. Outside the body of magnet they pass from north to South Pole

MULTIPLE CHOICE QUESTIONS

1. Following object is not attracted by magnet
   a) Iron
   b) Nickel
   c) Aluminum
   d) Steel
2. Magnetic length of magnet is ____ of geometric length of magnet
   a) 50%
   b) 71%
   c) 86%
   d) 100%
3. The line perpendicular to long axis of magnet is known as…
   a) Magnet meridian
   b) Equatorial line
   c) Central line
   d) None of the above

4. Identify the incorrect statement
   a) Strength North and South Poles of magnet are always are of equal strength
   b) Like poles repel and unlike poles attract each other
   c) Magnetic properties of magnet are not affected by temperature
   d) When magnet is free to move when suspended, it will stop in North-South direction
5. Outside the magnet, the direction of magnetic lines of force is from…
   a) South to North Pole
   b) North to South Pole
   c) South to South Pole
   d) North to North Pole
6. Following line is not perpendicular to magnetic meridian
   a) Axial line
   b) Equatorial line
   c) Both of the above
   d) None of the above
7. Magnetic force F between two magnets having magnetic strengths of m1 and m2, separated by distance r is given by following formula:
   a) $F = K m_1 m_2 / r$
   b) $F = K m_1 m_2 / r^2$
   c) $F = K m_1^2 m_2^2 / r$
   d) None of the above
8. Find out correct statement for magnetic lines of force
   a) They contract laterally
   b) They dilate longitudinally
   c) Both of the above
   d) None of the above
9. For lifting heavy iron bars which of the following magnets are used?
   a) Permanent magnets
   b) Temporary magnets
   c) Electromagnets
   d) All of the above

Answers:

1—c	4—c	7—b
2—c	5—b	8—d
3—b	6—b	9—c

Bibliography:
1. Singh J. Textbook of Electrotherapy, 1st ed. Jaypee Brothers Medical Publishers (P) Ltd. 2007.

Reference websites:
1. http://en.wikipedia.org
2. http://tutorskingdom.com

3. http://www.coolmagnetman.com
4. http://www.excellup.com
5. http://www.howmagnetswork.com
6. http://www.ndt-ed.org
7. http://www.school-for-champions.com
8. http://www.winnerscience.com

Electromagnetic Spectrum — Its uses and Governing Laws | 9

The electromagnetic spectrum is the range of all possible frequencies of electromagnetic radiation. The "electromagnetic spectrum" of an object is the characteristic distribution of electromagnetic radiation emitted or absorbed by that particular object.

TERMINOLOGIES

Wavelength (Fig. 9.1)

It is the spatial period of the wave means it is the distance over which the wave's shape repeats. It is usually determined by considering the distance between consecutive corresponding points of the same phase, such as crests, troughs, or zero crossings, and is a characteristic of traveling waves and standing waves, as well as other spatial wave patterns.
- **Unit**
  - Meter (m), centimeter (cm), etc.
- **Symbol**
  - λ (Lambda).

Frequency

It is the number of occurrences of a repeating event per unit time.
- **Unit**
  - Hertz (Hz) or Pulse per second (pps).
- **Symbol**
  - f.

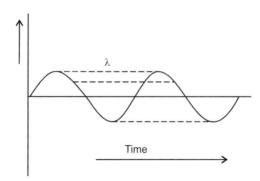

Fig. 9.1: Wavelength

Velocity

It is the speed at which wave propagates in the space.
For all electromagnetic waves it is 3×10^8 m/s
- **Unit**
  - Meter per second (m/s).
- **Symbol**
  - c.

Relationship Between Wavelength, Frequency and Velocity of Electromagnetic Waves

Formula

- $c = \lambda \times f$
- Where c is for velocity of electromagnetic waves in space (meter per second)
- λ is wavelength (meter)
- f is frequency (hertz).

The electromagnetic spectrum extends from low frequencies used for modern radio communication to gamma radiation at the high-frequency.

HISTORY

For a long time, light was the only believed a part of the electromagnetic spectrum.
- 1800 (William Herschel): The first discovery of electromagnetic waves other than light came, when he discovered infrared light. He noticed that the hottest temperature was beyond red. He theorized that there was 'light' that you could not see.
- 1801 (Johann Ritter): He worked at the other end of spectrum and noticed that there were 'chemical rays' that behaved similar to visible violet light rays. They were later renamed ultraviolet radiation.
- 1845 (Michael Faraday): He noticed that light responded to magnetic field so light was linked in electromagnetic spectrum for first time.
- 1860s (James Maxwell): He was studying electromagnetic field and realized that they traveled at around the speed of light.
- 1886 (Heinrich Hertz): He built an apparatus to generate and detect radio waves. He was able to observe that they traveled at the speed of light. In a later experiment he similarly produced and measured microwaves. These new waves paved the way for inventions such as the wireless telegraph and the radio.
- 1895 (Wilhelm Rontgen): He noticed a new type of radiation emitted during an experiment. He called these X-rays and found they were able to travel through parts of the human body but were reflected by bones. Before long many uses were found for them in the field of medicine.

- 1900 (Paul Villiard): He was studying radioactivity. He first thought they were particles similar to alpha and beta particles.
- 1910 (Ernest Rutherford): He measured their wave lengths and found that they were electromagnetic waves.

TYPES

The types of electromagnetic radiation are broadly classified into the following classes:

1. Gamma radiation
2. X-ray radiation
3. Ultraviolet radiation
4. Visible radiation
5. Infrared radiation
6. Microwave radiation
7. Radiowaves.

Note that some of the waves have not particular demarcation lines those differentiate them from other. In other words it can be said that some of portions of electromagnetic spectrum are occupied by more than one type of waves (Fig. 9.2).

Details about different waves including their characteristics, what are the uses and dangers are explained later.

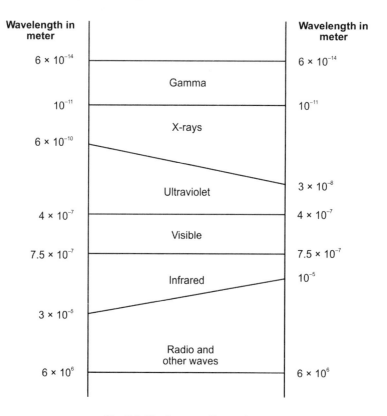

Fig. 9.2: Electromagnetic spectrum

LAWS GOVERNING ELECTROMAGNETIC SPECTRUM

REFLECTION (FIG. 9.3)

A "normal" is line drawn perpendicular to surface of medium at the point where an electromagnetic waves strikes. Angles of incidence and angle of reflection are measured between electromagnetic wave and normal. Reflection occurs when an electromagnetic wave encounters a medium which will not transmit it. In this case the ray is reflected back in the same plane such that the angle between the incident ray and normal will be equal to angle between the reflected ray and normal. The laws of reflection are employed in design of reflectors used for redirection of rays towards an appropriate target. If the incident angle is 0° (i.e. the radiation strikes the surface at right angle) the angle of reflection is also 0° (the incident ray, normal and reflected ray all coincide.

REFRACTION (FIG. 9.4)

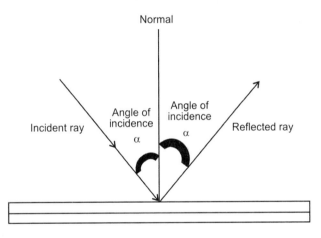

Fig.9.3: Reflection with normal, incident ray, reflected ray, angle of incidence, angle of reflection

Refraction occurs when electromagnetic rays are transmitted from one medium to another with an angle of incidence greater than zero. Rays with zero angle of incidence, i.e. striking the surface at right angles, continue in same straight line.

Refraction occurs depending on media involved and angle of incidence (Snell's law). When passing into a denser medium from less dense medium, the ray is refracted towards normal. When passing into less denser medium from high denser medium, ray is refracted away from normal. Refraction is important when using hydrotherapy as a form of treatment, as refraction will be making it difficult to identify exact posi-

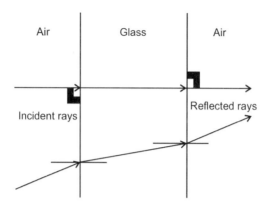

Fig.9.4: Refraction in two different mediums (air and glass) with angle of incidence 90° and other than 90°

tion of steps. Similar will be applicable while applying ultrasound with water as medium.

SNELL'S LAW OR SNELL–DESCARTES LAW OR LAW OF REFRACTION

This law explains formula to describe the relationship between the angles of incidence and angle of refraction, when light or other waves pass from one medium to another medium, such as water and glass. Although named after Dutch astronomer Willebroad Snellius (1580–1626), the law was first accurately described by the Arab scientist Ibn Sahl at Baghdad court.

According to Snell's law the ratio of sine values of the angles of incidence and refraction is equivalent to the ratio of phase velocities (Rate of phase of wave through which wave propagates in the space) in the two media, or equivalent to the inverse of ratio of the indices of refraction.
$\sin\theta_1 / \sin\theta_2 = v_1/v_2 = n_2/n_1$

Where θ_1 is angle of incidence and θ_2 is the angle of refraction. Angle is measured from the normal.

v_1 is the velocity of light in medium from which it is exiting and v_2 is velocity of light in medium where it is entering. (Meters per second)

n_1 is the refractive index of the first medium and n_2 is refractive index of second medium.

Refraction of light at the interface between two media of different refractive indexes have relation in such a way that $n_2 > n_1$. Since the velocity is lower in the second medium ($v_2 < v_1$), the angle of refraction θ_2 is less than the angle of incidence θ_1. This can be understood by one example. What happens if rays pass from air to water? Refractive indices of air and water are 1.00 and 1.33 respectively. So velocity of wave will be more in air than that in water as refractive index of air is less than that of water. So at interface, angle of incident ray is more than angle of refraction. So it can be understood that when ray passes from less dense medium, e.g. air to high dense medium, e.g. water it changes its direction and comes toward the normal. And when ray passes from denser medium to less dense medium, it changes its direction and goes away from normal.

ABSORPTION

When electromagnetic rays strike a new medium they may be absorbed and produce their effect. The proportion of rays absorbed depends on wavelength of rays, nature of medium and angle of incidence. Absorption is reciprocal of penetration, i.e. greater the penetration, lesser the absorption.

"Filter" is a medium which will absorb some electromagnetic waves while allowing others to pass. For example some window glass may allow visible light and infrared rays to pass but absorb ultraviolet rays. While water absorbs infrared and allow visible and ultraviolet rays to pass. X-rays are passing from soft tissues of human body but are absorbed by bones.

INVERSE SQUARE LAW

An inverse square law is stating that a specified physical quantity or strength at particular space is inversely proportional to the square of the distance from the source of that physical quantity. After the beam of electromagnetic waves emerges from point source, it will diverge as shown in Figure 9.5 so the intensity per unit area decreases.

- **Formula**
  - $I \alpha 1/d^2$
  - Where I is intensity or strength of electromagnetic waves at particular point of space.
  - d stands for distance between target point and source of wave.

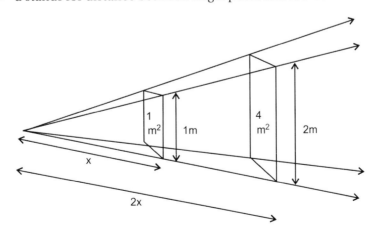

Fig. 9.5: Diversion of wave according inverse square law

The intensity of light or other linear waves radiating from a point source (energy per unit of area) is inversely proportional to the square of the distance from the source. So an object at double distance, receives only one-quarter the energy.

For example, the intensity of radiation from the Sun is 9126 Watts per square meter at the distance of Mercury planet but only 1367 Watts per square meter at the distance of earth. Distance between earth and sun is three times of that between mercury and sun. An approximate threefold increase in distance results in an approximate nine fold decreases in intensity of radiation.

If physiotherapist is giving microwave diathermy to patient, just assume that distance between microwave and patient is 10 cm and intensity set is 20 Watts. Now if physiotherapist doubles the distance, i.e. 20 cm, so the part of patient treated will get only 1/4th of intensity than that of first placement. So physiotherapist will have to increase intensity four times, i.e. 80 V to give same effect as first placement. If distance is kept 30 cm now, intensity should be increased nine times than that of first placement, i.e. 180 Watts.

Inverse square law is also applicable for gravity and electronics. For example of gravity, forces between two objects will be directly proportional to product of mass of two objects and inversely proportional to square of distance between them. Like this way only, in electronics, attraction between two charges will be directly proportional to product of charge and inversely proportional to square of distance between them.

ATTENUATION OR EXTINCTION

It is the gradual loss in intensity of any kind of electromagnetic waves through a medium. For instance, sunlight is attenuated by dark glasses, X-rays are attenuated by lead, and light and sound are attenuated by water.

In electrical engineering and telecommunications, attenuation affects the propagation of waves and signals in electrical circuits, in optical fibers and also in air.

Factors Affecting Attenuation

Absorption

Waves will be absorbed in the tissue and converted to heat at that point. This contributes the thermal effect any wave, e.g. ultrasound.

Scatter (to spread)

This occurs when normally cylindrical beam is deflected from path by reflection at interfaces, bubbles or particles in the pathway.

The overall effect of these two is such that beam is reduced in intensity the deeper it passes. This gives rise to expression of "half value distance" which means that distance from source of wave or depth of tissues where beam intensity becomes half as compared to intensity at surface. Half value distance for ultrasound of 1 MHz and 3 MHz is 4 cm and 2.5 cm respectively.

Measurement

Attenuation is usually measured in units of decibels per unit length of medium (dB/cm, dB/km, etc.) and is represented by the attenuation coefficient of the medium.

Attenuation Coefficient

It is used to quantify different media according to how strongly the transmitted ultrasound amplitude decreases.

Attenuation of Different Waves

Ultrasound

Attenuation in ultrasound is the reduction in amplitude of the ultrasound beam as it passes more distance. Consideration for attenuation effects in ultrasound is important because reduced signal amplitude can affect the quality of the image produced in ultrasonography for any soft tissue injury or fetal examination for diagnostic purpose. By knowing the attenuation that an ultrasound beam experiences traveling through a medium, one can adjust the amplitude to compensate for any loss of energy to provide particular intensity to target tissue. Even when ultrasound is used for treatment purpose in physiotherapy department, consideration should be taken that for deeper tissues pathology more intensity of ultrasound is required to compensate for reduction in intensity due to attenuation.

Earthquakes

Attenuation also occurs in earthquakes; when the seismic waves move farther away from the epicenter. As distance increases from epicenters intensity of waves will decreases.

Sunlight

When the sunlight reaches the water surface, it is attenuated by the water, and the intensity of light decreases with water depth.

Electromagnetic

Attenuation decreases the intensity of electromagnetic radiation due to absorption or scattering of photons. Attenuation is an important consideration in the modern world of wireless telecommunication. Attenuation limits the range of radio signals and is affected by the materials a signal must travel through (e.g. air, wood, concrete, rain).

Remember that attenuation does not include the decrease in intensity due inverse square law. Therefore, calculation of the total change in intensity involves both the inverse square law and an estimation of attenuation over the path.

COSINE LAW (FIG. 9.6) OR LAMBERT-COSINE LAW

It explains relationship between angle of incidence and absorption of waves. Amount of absorption of rays depends on cosine value of angle of incidence, i.e. angle between normal and incident ray.

So larger the angle at which the rays strike at the body surface, lesser will be the absorption. As per shown in the diagram at 90° of angle of incident rays with normal, no rays will be absorbed (cos90° = 0). At 0° of angle between incident rays with normal absorption is maximum (cos0°= 1).

At angle of 45° between incident rays with normal 70% waves will be absorbed as cos45° = 0.7.

Cosine law is very useful in physiotherapy department in clinical implication. To maximize effect of any radiation (e.g. infrared, ultraviolet, etc.) reflector should be arranged perpendicular to body surface, i.e. angle of incidence 0° from normal, so that absorption will be maximum and treatment will be more effective by avoiding waste of energy.

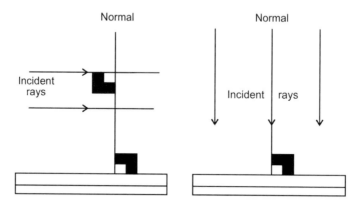

Fig. 9.6: Incident rays at 90° of angle of incidence and at 0° of angle of incidence for Cosine law respectively

GROTHUS LAW OR GROTTHUS DRAPER LAW

It was explained first by Grothus in 1820. It states that electromagnetic waves must be absorbed to produce their effect. The effect will be produced at that point at which the rays are absorbed. This means that if waves are absorbed by tissue, then only they can produce their effect but if they are reflected from tissue surface, they will not produce any effect.

USES OF ELECTROMAGNETIC WAVES

Uses of different electromagnetic waves are explained in Table 9.1.

Table 9.1: Characteristics uses and dangers of different electromagnetic waves

Name of wave	Characteristics	Uses	Dangers
Gamma radiation	• Given off by stars and some radioactive substances • Extremely high frequency waves carry a large amount of energy • Pass through most materials so are quite difficult to stop	• They are used to kill cancer cells This is called "Radiotherapy", and works because cancer cells cannot repair themselves like normal cells • Gamma rays are used to sterilize food • Gamma rays are also used to sterilize medical and surgical equipments	• Gamma rays can cause cancers if not controlled • They cause mutations in growing tissues, damage to fetus is possible
X-ray radiation	• Given off by stars • High frequency waves and carry a lot of energy • Pass through most substances	• X-rays are used by radiologists to see inside human body as they pass easily through soft tissues, but not through bones • Barium meal by patient absorbs X-rays, and so the patient's intestines will show clearly by X-ray • X-rays are also used in airport security service to look for suspected things in luggage • They are also used by astronomers. Many objects in the universe emit X-rays, which we can detect using suitable radiotelescopes • Lower energy X-rays cannot pass through tissues as easily, so used to scan soft areas such as the brain by CT scan	• Over dose of X-rays can cause cell damage and cancers

Contd...

Contd...

Name of wave	Characteristics	Uses	Dangers
Ultraviolet radiation (UVR)	• They are made by special lamps. The lamp gives off invisible ultraviolet and visible violet light • It is given off by the Sun in large quantities • The UV attracts insects so they do not land on food	• Used for hardening of dental filling • They are also used for detecting invisible writing in currency notes • They are used to sterilize surgical equipments and the air in operation theaters • They are used to sterilize food products • Controlled doses of ultraviolet rays cause the body to produce vitamin D • Physiotherapists use UVR lamp for different skin treatment conditions	• Large doses of UVR can damage the retina. Sunglass can be used to protect our eyes from damage from ultraviolet rays emitted from sun and definitely to look handsome! • Overdoses or UVR cause sunburn and even skin cancer (Fortunately, the ozone layer in the Earth's atmosphere screens us from most of the UV given off by the Sun
Visible radiation	• They are given off by anything that's hot enough to glow • Sun gives light • White light is actually made up of a whole mixture of colors • If light passes through a glass prism, by dispersion rainbow is seen	• We use light to see objects • Light waves can also be made using a laser. Lasers are also used in laser printers, and in aircraft weapon aiming systems	• Too much light can give reversible damage to the retina in your eye if staring continuously at sun or any bright object
Infrared radiation (IR)	• Given off by any hot objects, stars, lamps, flames and anything else that's warm - including human body	• Some of the images for weather information are taken using IR cameras	• Overheating

Contd...

Electromagnetic Spectrum —Its uses and Governing Laws

Contd...

Name of wave	Characteristics	Uses	Dangers
	(The detector will be used to identify presence of infrared because with naked eyes they cannot be seen)	• They are used remote controls for TVs and video recorders • IR is also used for short-range communications • Night sights for weapons sometimes use a sensitive IR detector • Some of the camera uses IR so that vision objects look brighter • To reveal secret writings on ancient walls • In green house, to keep plants warm • Physiotherapists use IR lamps for increasing blood circulation	
Microwave radiation	• In a mobile phone, they are made by a transmitter chip and an antenna • In a microwave oven they are made by a "magnetron" • Also given off by stars	• They are used to cook many types of food • Mobile phones use microwaves • They are also used by fixed traffic speed cameras, radar which is used by aircraft, ships and weather forecasters	• Prolonged exposure can lead to cataracts • Microwaves from mobile phones can affect brain
Radio-waves	• Given off by stars, sparks and lightning, various types of transmitter	• They are used mainly for communications	• Large doses cause cancer, leukemia and other disorders

MULTIPLE CHOICE QUESTIONS

1. _____ law states that waves must be absorbed to produce their effects
   a) Grothus law
   b) Cosine law
   c) Inverse square law
   d) Ohm's law
2. Select the true one
   a) Angle of incidence and angle of reflection are same
   b) Angle of incidence is more than angle of reflection
   c) Angle of incidence is less than angle of reflection
   d) No relation exists between angle of incidence and angle of reflection
3. If distance between source of rays and reflecting surface increases, intensity of rays on surface
   a) Increases
   b) Does not change
   c) Decreases
   d) Increases or decreases
4. Intensity of waves will decrease as it passes more distance due to their spreading is called…
   a) Attenuation
   b) Cosine law
   c) Inverse square law
   d) Ohm's law
5. If surface does not allow the ray to pass _____ will occur
   a) Refraction
   b) Penetration
   c) Absorption
   d) Reflection
6. If distance between source and surface doubles intensity will…
   a) Becomes 1/4th of original intensity
   b) Becomes half of original intensity
   c) Remains as it is
   d) Becomes double of original intensity
7. Angle of incidence is…
   a) Angle between incident ray and reflected ray
   b) Angle between incident ray and surface
   c) Angle between incident ray and normal
   d) None of the above
8. If rays strike at 30° with the surface amount of rays absorbed is…
   a) 100%
   b) 85%
   c) 50%
   d) 0%

9. Select the correct one...
   a) Velocity = wavelength / frequency
   b) Velocity = wavelength × frequency
   c) Wavelength = velocity × frequency
   d) Frequency = velocity × wavelength
10. When ray enters from less denser medium to high denser medium...
    a) Refracts away from the normal
    b) Refracts towards the normal
    c) Does not refract
    d) Refracts with 45° with the normal
11. Refraction depends on ...
    a) Media involved and angle of incidence
    b) Media involved and angle of refraction
    c) Angle of refraction and angle of incidence
    d) Media involved, angle of incidence and angle of refraction
12. Select the correct one...
    a) Absorption increases if penetration is more
    b) Absorption decreases if penetration is more
    c) Absorption decreases if penetration if less
    d) None of the above
13. In design of reflector following principle is used...
    a) Law of reflection
    b) Law of refraction
    c) Cosine law
    d) Grothus law
14. For increasing blood circulation which of the following electromagnetic waves are used?
    a) Infrared
    b) Ultraviolet rays
    c) Gamma rays
    d) X-rays
15. Range of wavelength of electromagnetic spectrum ranges from
    a) 6×10^{-18}m to 6×10^{4}m
    b) 6×10^{-16}m to 6×10^{4}m
    c) 6×10^{-14}m to 6×10^{6}m
    d) 6×10^{-12}m to 6×10^{8}m
16. Following rays are used not used for treatment purpose...
    a) X-rays
    b) Infrared rays
    c) Ultraviolet rays
    d) None of the above
17. Following rays are used for treatment of skin conditions...
    a) Gamma rays
    b) Infrared rays
    c) Ultraviolet rays
    d) X-rays

18. Following factor is not responsible for attenuation...
    a) Absorption
    b) Scatter
    c) Angle of incidence
    d) None of the above
19. Sign of attenuation is...
    a) c
    b) λ
    c) f
    d) None of the above

Answers:

1—a	6—a	11—a	16—a
2—a	7—c	12—b	17—c
3—c	8—c	13—a	18—c
4—c	9—b	14—a	19—d
5—d	10—b	15—c	

Bibliography:
1. Singh J. Textbook of Electrotherapy, 1st ed. Jaypee Brothers Medical Publishers (P) Ltd, 2007.

Reference websites:
1. http://en.wikipedia.org
2. http://hyperphysics.phy-astr.gsu.edu
3. http://interactagram.com

Electromagnetic Induction 10

It is the production of an electric current across a coil when it moves through a magnetic field of magnet. Current is produced in this way even though magnet and coil are not in contact. It underlies the operation of generators, transformers, induction motors and electric motors.

- **Inventor**
  - Michael Faraday (1800).
- **Magnetic flux**
  - Magnetic flux is the strength of the magnetic field passing through any given surface area.
- **Measuremen**
  - By flux meter
- **Unit**
  - Weber (Wb) or volt-seconds
- **Symbol**
  - Φ
- **Calculation**
  - Electromotive force (EMF) produced around a closed path is proportional to the rate of change of the magnetic flux through any surface that is given by Faraday's second law. It means that an electric current will be induced in any close circuit when the magnetic flux around coil changes. This applies whether the field itself changes in strength because of magnetic movement or the conductor is moved through it or both with unequal speed.
- *Factors for electromagnetic induction are*
  - A coil of conducting material
  - Magnetic lines of force from magnet
  - Relative movement between these two.

In a coil of conductor magnetic lines of force emerging from magnet will induce movement of free electron. This electron will stimulate movement of another adjacent one. In this way, flow of electron that is current is produced in the coil. Magnetic lines of force produced around one coil due to electric current passing through it, can also produce electromotive force in another coil by the same principle. This means that magnetic lines of force are able to produce electromotive force in coil irrespective of its source whether it is magnet or another coil. But there must be change in magnetic lines of force with time. To get this, relative movement between coil and magnetic lines of force is must.

For current carrying coil, alternating current will be more useful. Alternating current is continuously changing its intensity and direction so

that it will lead to change in magnetic lines force of magnetic field for current carrying coil. In dynamo, conductor is rotated in the magnetic field and electric current will be produced. This makes it possible for the light of vehicle to glow.
- **Unit**
  - Henry
- **Symbol**
  - L

SELF-INDUCTION

When the current from battery passes from any circuit, Electromotive Force (EMF) will be produced in the same circuit. This induced EMF will oppose any change in strength of current flowing through the circuit. This is known as self-induction. Because of this it is also known as inertia of electricity.

Explanation

When current starts to flow from any circuit, current will rise from zero to maximum. This change in strength of current will also change magnetic flux linked to circuit which induces EMF in the circuit. This EMF will induce current in same circuit. This induced current will flow in direction opposite to original current and will oppose increase in original current from zero to maximum. This happens as per Lenz's law.

When current begins to fall from maximum to zero, this change in strength of current will also change magnetic flux linked to circuit which induces EMF in the circuit. This EMF will also induce current the same circuit. This induced current will flow in direction of original current and will oppose decrease in original current from maximum to zero. This phenomenon also follows Lenz's law.
- **Unit**
  - Henry

When current changes at the rate of one ampere per second with induced EMF of one volt in the circuit, self-induction is said of 1 Henry.

MUTUAL INDUCTION

When two circuits are nearer to each other, current in one circuit will produce EMF around it. This EMF will oppose any change in strength of current in another circuit.

Explanation

For example, two circuits are very nearer to each other. When first circuit is attached with battery, current will start to flow from zero to maximum. During this time, because of change in strength of current in first circuit, magnetic flux linked with this circuit will also change. This will induce EMF in second circuit. This induced EMF will produce current in second

Electromagnetic Induction

circuit in such a way that it will oppose the increase in the current flowing through first circuit according to Lenz's law.

Like wise, in above example, when current starts to fall in first circuit from maximum to zero. During this time, because of change in strength of current in first circuit, magnetic flux linked with this circuit will also change. This will induce EMF in second circuit. This induced EMF will produce current in second circuit in such a way that it will oppose the decrease in the current through first circuit according to Lenz's law.

- **Unit**
  - Mutual inductance of two circuits is one Henry when a current changes at the rate of one ampere in one second in first coil with induced EMF of one volt in second coil.
- *Factors affecting mutual inductance*
  - Size of coil
  - Shape of coil
  - Number of turns
  - Material of coil
  - Distance between two circuits
  - Relative arrangement of two circuits.

INDUCTIVE REACTANCE

Inductive reactance is an opposition to the alteration of electric current in circuit.

- **Formula:**
  - $X_L = 2fL$
  - Where, X_L is inductive reactance (ohm)
  - f is frequency (hertz)
  - L is inductance (henry).

FARADAY'S LAW

They are two in number. They are explained below:

First Law

When amount of magnetic flux associated with a circuit changes, an electromotive force is induced in the circuit. Induced electromotive force lasts as long as change in magnetic flux exists. When any coil comes in magnetic field of any conductor, effect of magnetic field will be on that coil. When magnet moves towards the coil, magnetic lines of force linked with coil increases and when magnet moves away from the coil, magnetic lines of force linked with coil decreases. In both cases current will be produced in the coil. In this case, galvanometer shows deflection. When magnetic lines of force remain constant, i.e. if magnetic field and coil are relatively stationary, current will not be produced in the coil. In this case galvanometer does not show any deflection.

Second Law

The magnitude of induced electromotive force in circuit is directly proportional to the rate of change of magnetic flux linked with a circuit. When magnet is moving faster, magnetic flux will change faster and galvanometer deflection will be more. But when magnet moves slower, magnetic flux will change slower and galvanometer deflection will be less.
- Formula for second law of Faraday is as below
  - $E = -d\Phi/dt$
  - Where E is electromotive force
  - $d\Phi$ is change in magnetic flux
  - dt is change in time
  - Negative sign is because induced electromotive force always opposes any change in magnetic flux linked with coil.

LENZ'S LAW

It is a common way of understanding how electric circuits obey Newton's third law and the conservation of the energy. Lenz's law is named after Heinrich Lenz.
- According to this law
  - An induced electromotive force always gives rise to a current whose magnetic field opposes the original change in magnetic flux.
  - When magnet is moved towards the coil, electric current is induced in the coil. The direction of current will be such that it opposes the push of magnet towards coil. Magnetic field which is produced due to induced current will exhibit same magnetic pole towards moving magnet. So according to the principle, "like poles repel each other." movement of magnet towards coil will be opposed. The same will be applicable if coil is moving towards the magnet.
  - When magnet is moved away from coil, at that time also electric current is induced in the coil. The direction of electric current is such that it will oppose the pull of magnet from coil. Magnetic field which is produced due to induced current will exhibit opposite magnetic pole towards moving magnet. So according to the principle, "unlike poles attract each other." Movement of magnet away from coil will be opposed. The same will be applicable if coil is moving away from magnet.

EDDY CURRENT (FIG. 10.1) OR FOUCAULT CURRENT

This phenomenon was first observed by Focault in 1895. This is electric current induced in conductors when a conductor is exposed to a changing magnetic field. This happens due to relative motion of the magnetic field and coil. It may also occur due to variations in the field with time. This can cause a circulating flow of electrons, or current, within the con-

Electromagnetic Induction

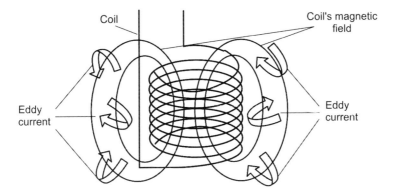

Fig. 10.1: Eddy current

ductor. These circulating eddies of current induce magnetic fields. These fields can cause repulsive, attractive, propulsive and drag effects.

The amount of induced current and strength of induced filed depends on:
- Applied magnetic field
- Electrical conductivity of the conductor
- Rate of field changes.

The term eddy current comes from analogous currents seen in water when dragging an oar. Somewhat analogously, eddy currents can take time to build up and can persist for very short times in conductors due to their inductance.

Eddy currents, like all electric currents, generate heat as well as electromagnetic forces. Eddy currents can also have undesirable effects, for instance power loss in transformers.

Uses of Eddy Current

- Metal detectors
- Electricity meters (electromechanical induction meters)
- Induction heating
- Traffic detection systems
- Vending machines (detection of coins)
- Mechanical speedometers
- Safety hazard and defect detection applications, etc.

MULTIPLE CHOICE QUESTIONS

1. When current is produced in the coil because of change in magnetic field of the conductor, it is known as…
   a) Electric induction
   b) Magnetic induction
   c) Electromagnetic induction
   d) None of the above

2. _____ was pioneer in the invention of electromagnetic induction
   a) Maxwell
   b) Hertz
   c) Faraday
   d) Joule
3. Electromagnetic induction is very useful for the operation of following instruments.
   a) Electric generators
   b) Transformers
   c) Electric motors
   d) All of the above
4. Following type of current is used more for electromagnetic induction in various apparatus.
   a) Direct current
   b) Pulsed current
   c) Alternating current
   d) Both of a and c
5. Unit of electromagnetic induction is...
   a) Faraday
   b) Henry
   c) Coulomb
   d) Joule
6. Direction of induced electromotive force will follow the direction according to which law?
   a) Faraday's law
   b) Lenz's law
   c) Joule's law
   d) Coulomb's law
7. If current passes through any circuit produce electromotive force in the same circuit, it is known as...
   a) Self-induction
   b) Mutual induction
   c) Any of the above
   d) None of the above
8. Pick up the correct statement.
   a) An induced electromotive force always gives rise to a current whose magnetic field opposes the original change in magnetic flux
   b) An induced electromotive force always gives rise to a current whose magnetic field favors the original change in magnetic flux
   c) An induced electromotive force always resists the current whose magnetic field opposes the original change in magnetic flux
   d) None of the above

9. Eddy current was invented by…
    a) Faraday
    b) Joule
    c) Coulomb
    d) Foucault

Answers:

1—c	4—c	7—a
2—c	5—b	8—a
3—d	6—b	9—d

Bibliography:

1. Singh J. Textbook of Electrotherapy, 1st ed. Jaypee Brothers Medical Publishers (P) Ltd, 2007.

Reference websites:

1. http://en.wikipedia.org
2. http://hyperphysics.phy-astr.gsu.edu
3. http://nuvvo.com
4. http://www.britannica.com
5. http://www.launc.tased.edu.au
6. http://www.physicshandbook.com
7. http://www.physlink.com
8. http://www.wisegeek.com

Transformer {11}

A transformer is a device that transfers electrical energy from one circuit to another through inductively coupled conductor—The transformer's coils. A varying current in the first or primary circuit creates a varying magnetic flux in the transformer's core. This varying magnetic field induces a varying electromotive force in the secondary circuit. This effect is called inductive coupling.

If a load is connected to the secondary, current will flow in the secondary circuit, and electrical energy will be transferred from the primary circuit through the transformer to the load.

Size

Transformers range in size from a thumbnail-sized coupling transformer hidden inside a stage microphone to huge units weighing hundreds of tons used to interconnect power grids. All operate on the same basic principles, although the range of designs is wide. Nowadays new technology has minimized the need for transformers in some electronic circuits, transformers are still found in almost all electronic devices designed for household voltage.

■ Static transformer

- **Principle**
  - Electromagnetic (mutual) induction.

Construction

It is made up of two coils of insulated wired wound on a soft iron core. These two coils may be wound on top of one another or on opposite sides of the frame. One is known as primary coil and another is known as secondary coil.

Mechanism of Work

When alternating current is passed through primary coil, magnetic field will be produced and continuously changing associated with this coil. This magnetic field lines cuts secondary coil and electromotive force is also produced in secondary coil.

Types According to Functions

Its function is to alter the voltage of alternative current. It depends on number of turns of in both coils relative to each other. Details are as follows:

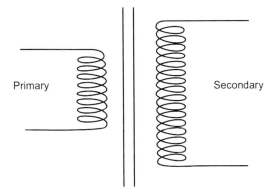

Fig.11.1: Step up transformer

Even ratio transformer

In this number of turns will be same in both coils. So voltage in each coil will be same.

Step up transformer (Fig. 11.1)

In this type, number of turns will be more in secondary coil than those of primary coil. For example, number of turns is 50 in primary coil, and voltage is 60 volts. So if number of turns is 100 that are double in secondary coil, voltage will also be double 120 volts in secondary coil.

Step down transformer (Fig. 11.2)

In this type, number of turns will be less in secondary coil than those of primary coil. For example, number of turns is 50 in primary coil, and voltage is 60 volts. So if number of turns are 25 that is half in secondary coil, voltage will also be half 30 volt in secondary coil.

In an ideal transformer, the induced voltage in the secondary circuit (V_2) is in proportion to the primary voltage (V_1) and is given by the ratio

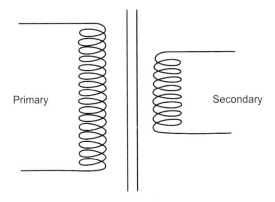

Fig. 11.2 Step down transformer

of the number of turns in the secondary (N_2) to the number of turns in the primary (N_1) as follows:

$N_1/N_2 = V_1/V_2$

By appropriate selection of the ratio of turns, a transformer thus enables an alternating voltage to be "stepped up" by making N_2 greater than N_1, or "stepped down" by making N_2 less than N_1.

Electrical power is same in both primary and secondary coils. Electrical power has unit Watts. And as per formula, Electrical power = Potential difference × Current (Watts = volts × amperes). $V_1I_1 = V_2I_2$

▋ Variable transformer (Fig. 11.3)

It has also two coils: One primary and one secondary. Mostly primary coil has facility to change its length so that numbers of turns can be altered. By moving contact points on primary coil, length of it can be changed. By this way, very rough control over voltage can be achieved.

▋ Autotransformer

It is an electrical transformer with only one winding. The auto prefix refers to the single coil acting on itself but not any automatic mechanism. In an autotransformer primary and secondary coils are portions of the same winding. It can be smaller, lighter and cheaper than a standard transformer. Autotransformers are often used to step up or down voltages. It is used in circuit of ultraviolet lamps. They operate on time-varying magnetic fields and so will not function with direct current.

- Uses

  They are as follows:
  - To render the current earth free: In most of the electronic instruments current is carried by live wire to the load. Sometimes if live wire comes in contact with earth, current starts to enter into earth as it is very good conductor of electricity. If any person comes in

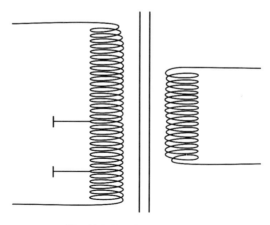

Fig. 11.3: Variable transformer

Transformer   **89**

pathway of current, he or she may receive earth shock. To prevent this, if static transformer is used, secondary coil has no connection with earth. Electron flow produced in secondary coil due to changes in magnetic field of primary coil will not be able to enter into earth and earth shock is not possible.
- For welding instruments
- In regulators of electronic instruments
- For transmission of alternating current
- High voltage discharge lamp
- For door bell installation.

- *Energy loss from transformer*
  - It is common from metals like copper and iron because of heat production according to Joule's law, leakage of magnetic flux, noise from the circuit.

MULTIPLE CHOICE QUESTIONS

1. Functions of transformer is
   a) To alter the voltage
   b) To alter the power
   c) Both of the above
   d) None of the above
2. In a step up transformer voltage is increased upto 100 volts from 50 volts. The current in primary circuit is 1 ampere. So current in secondary circuit will be…
   a) 0.5 ampere
   b) 1 ampere
   c) 2 ampere
   d) 4 ampere
3. In a step down transformer, voltage is decreased 250 volts from 500 volts. The current in secondary circuit is 1 ampere, so the current in primary circuit will be…
   a) 0.5 ampere
   b) 1 ampere
   c) 2 ampere
   d) 4 ampere
4. In a step up transformer, voltage is stepped up from 100 volts to 200 volts. The amount of current in primary circuit is 1 ampere. What will be the power in secondary circuit?
   a) 50 watt
   b) 100 watt
   c) 200 watt
   d) 400 watt
5. Transformer works on the principle of
   a) Electromagnetic radiation
   b) Electromagnetic induction
   c) Both of the above
   d) None of the above

6. In a step down transformer
   a) The number of turns will be same in both primary and secondary circuit
   b) The number of turns will be more in primary circuit than those of secondary circuit
   c) The number of turns will be less in primary circuit than those of secondary circuit
   d) The number of turns cannot be commented from type of transformer
7. Autotransformer is used in…
   a) Electrical stimulation machine
   b) Ultraviolet lamps
   c) Infrared lamps
   d) All of the above

Answers:
1—a   4—b   7—b
2—a   5—b
3—a   6—b

Bibliography:
1. Clayton's Electrotherapy. AITBS Publishers, 2002.

Reference websites:
1. http://en.wikipedia.org

Transistor | 12

Introduction

It is a device made up of semiconductor that is used to control the amount of current or voltage or used for amplification/modulation or switching of an electronic signal (Fig. 12.1).

History

It was invented by John Bardeen, Walter Brattain and William Shockley at the Bell laboratories in 1945. Today's computers contain millions of microscopic transistors.

Before the invention of transistors, digital circuits were made up of vacuum tubes having following disadvantages:
- They were much larger
- They required more electricity
- They produced more heat
- Repeated failures.

Without transistor, revolution of computer was almost impossible. Following its development, the transistor revolutionized the whole field of electronics.

Uses

- The transistor is the part of all microchips, including Central Processing Unit, and is what creates the binary 0's and 1's your computer use to communicate. In first computer, there were near about 2000 transistors.

Fig. 12.1: Transistors

- They are used for basic component of integrated circuit in radios, calculators and other things. A transistor controls a large electrical output signal with changes in small input signal. This is analogous to the small amount of effort required to open a tap to release a large flow of water. Since a large amount of current can be controlled by a small amount of current, a transistor acts as an amplifier.
- A transistor acts as a switch which can open and close many times per second.
- Transistors only make it possible to operate a loudspeaker, which makes sounds much louder than the person's voice.

Types

Junction transistors
- NPN transistor
- PNP transistor.

Field effect transistors
- Junction Field Effect Transistor (JFET)
- Metal Oxide Semiconductor Field Effect Transistor (MOSFET).

Junction transistors

The junction transistor consists of a layer of one type of semiconductor material between two layers of another type of semiconductor materials. If the middle layer is of P-type, the other layers must be N-type. This type is known as NPN transistor (Fig. 12.2). One outer layer is emitter and the other outer layer is collector. The middle layer is the base. The places between any two layers where they join are known as junctions.

In PNP junction transistors (Fig. 12.3), the emitter and collector are of P-type semiconductor material and the base is of N-type semiconductor materials. The one difference is existing in NPN transistor and PNP transistor.

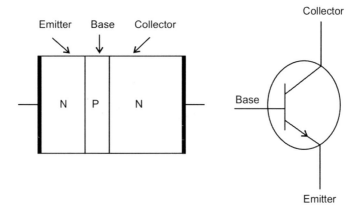

Fig. 12.2: NPN transistor with its electronic symbol

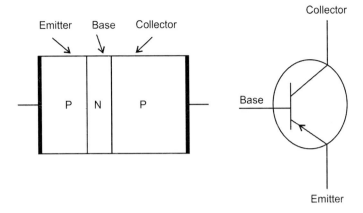

Fig.12.3: PNP transistor with its electronic symbol

In NPN transistor flow of current is controlled by electron while in PNP transistor flow of current is controlled by positive holes (Fig. 12.3).

Field Effect Transistors

It is made up of two layers of semiconductor materials, one is placed above another. The layer from which electricity flows is known as channel. And the layer from which voltage is connected to load is known as gate. The gate controls the intensity of the current coming from channel.

Mechanism of Work

A bipolar junction transistor has three terminals—Base, collector, and emitter corresponding to the three semiconductor layers of the transistor. The weak input current is applied to the inner (base) layer. When there is a small change in the current or voltage at the inner semiconductor layer (base), a rapid and far larger change in current takes place throughout the whole transistor (Fig. 12.4).

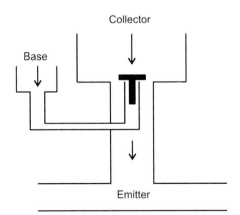

Fig. 12.4: Diagram to understand function of transistor

From above mentioned diagram it can be understood that when water comes from base, it will lift the plunger so water can come from collector to emitter. If water is not coming from base, plunger will block the water from collector to emitter. Very small amount of water from base, will allow large amount of water to enter from collector to emitter. In this way, a small input current of electricity to the base leads to a large flow of electricity from the collector to the emitter.

MULTIPLE CHOICE QUESTIONS

1. Transistor is made up of…
   a) Conductors
   b) Semiconductors
   c) Insulators
   d) Any of the above
2. Identify which is not part of transistor
   a) Base
   b) Emitter
   c) Collector
   d) None of the above
3. To get function of transistor, which area is selected for giving current?
   a) Base
   b) Emitter
   c) Collector
   d) None of the above
4. Transistors made up of three semiconductor materials with arrangement of outer two layers are of n-type and middle layer are of p-type, is known as…
   a) PNP transistor
   b) NPN transistor
   c) JFET transistor
   d) MOSFET transistor

Answers:
1—b   3—a
2—d   4—b

Reference websites:
1. http://en.wikipedia.org
2. http://www.computerhope.com
3. http://www.physlink.com
4. http://www.reuk.co.uk
5. http://www.webopedia.com

Capacitor or Condenser   13

- **Type**
  - Passive
- **Inventor**
  - Ewald Georg von Kleist (October 1745).
- **Electronic symbol**
  - ─┤├─

Construction (Fig. 13.1)

It is a device for storing an electric charge. It consists of two conductors separated by a nonconductive region called dielectric. Capacitor plates are made up from mainly aluminum, tantalum, silver, brass and liquid mercury, etc.

Dielectric

It is poor conductor of electricity, but supporter of electrostatic field. Most commonly dielectric materials are solid, e.g. ceramic, mica, glass, plastics, and metals oxides. Very rarely liquids and gases are used as dielectrics, e.g. distil water, dry air. Dielectric must be wasting minimal energy in the form of heat.

Capacitance

It is measurement of capacity of capacitor. It is ability of conductor to hold the electric charge.

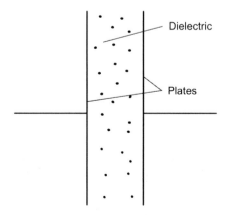

Fig. 13.1: Construction of capacitor

- **Unit**
  - Farad

  1 farad is the capacity of a conductor that is charged to a potential of 1 volt by 1 coulomb of electric charge. For practical purpose the microfarad (10^{-6} farad) is used.

- **Formula**
  - C = Q/V .... (1)
  - Where C is capacitance of capacitor (farad)
  - Q is electric charge (coulomb)
  - V is potential difference (volt).

The capacitance of capacitor depends on amount of electric charge with which it is charged and potential difference between two plates of capacitor.

$Q = CV$ if V is constant $C \alpha Q$

If a capacitor has large capacitance for given potential difference between two plates of capacitor, amount electric charge stored by capacitor is also more. The capacitance is directly proportional to electric charge.

$Q = CV$ if Q is constant so CV is constant $C \alpha 1/V$

If a capacitor has large capacitance then for a given quantity of electrons only a relatively small potential difference will be developed. The capacitance is inversely proportional to potential difference.

The capacitance of a capacitor is affected by following factors:
- Size of the plates
- Material of the plates
- Width of the dielectric
- Material of the dielectric.

Electric Field of Capacitor (Figs 13.2a and b)

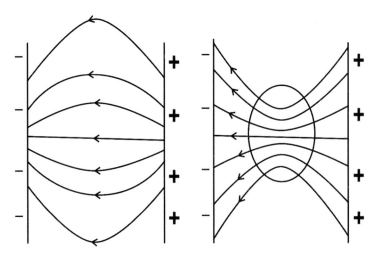

Fig.13.2a: Spread of electric field between plates of capacitor

Fig.13.2b: Spread of electric field passing from material of high dielectric constant

- If the plates are given opposite static electrical charges, the electric lines of force pass from one plate to another plate, from positive polarity to negative polarity.
- The electric filed between the plates of a charged capacitor consists of electric lines of force which tend to take the shortest possible pathway between the plates. However they repel one another so will occupy more area.
- They tend to pass more easily through some materials than others. They will travel through human body also.
- Concentration of this field depends on shape of surface of part where they enter.
- In given diagrams we can see effect of electric field on molecules. If there is no electric field between two plates of capacitor, moleculs will be under their natural structure. But if they come under influence of electric field between two plates of capacitor, nucleus is attracted towards negative polarity of plate and electrons with their orbits will be attracted towards positive polarity of plate of capacitor (Fig. 13.3).

Types of Capacitor

Depending on shape

Parallel plate capacitors: They are capacitors in which conductors used are parallel plates.

Spherical conductors: They are capacitors in which spherical conductors are used.

Cylindrical conductors: They are capacitors in which conductors used are of cylindrical type.

Depending on variability of capacitance

Fixed capacitor: They are made up of interleaved metal plates with air

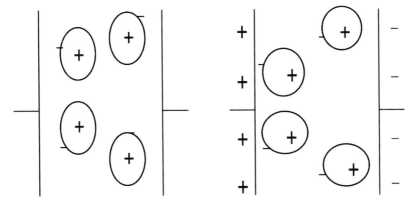

Fig. 13.3: Distortion of molecules under the influence of electric charge

as the dielectric. In this type of capacitors, capacitance of capacitor will be constant or fixed.

Variable capacitor: They consist of two sets of plates in such a way that one set of plates can be moved relative to each other. In instruments it will be controlled by tuning knob.

Uses of variable capacitor
- Radio sets
- Short wave diathermy machines.

Varying capacitance allows a circuit to be tuned to match the frequency of another oscillating circuit, thereby facilitating maximum transfer of energy between the two circuits.

Grouping of capacitors
1. Capacitors in series
2. Capacitors in parallel.

Capacitors in series (Fig. 13.4)

If capacitors are connected in series, the total voltage developed on them is sum of the voltage on each of them.
- $V = V1 + V2 + V3 + V4$ ....(2)

From formula (1)
- $C = Q / V$
- $1 / C = V / Q$

From formula (2)
- $V = V1 + V2 + V3 + V4$
- So, $1 / C = (V1+V2+V3+V4) / Q$
- $1 / C = V1/Q + V2/Q + V3/Q + V4/Q$

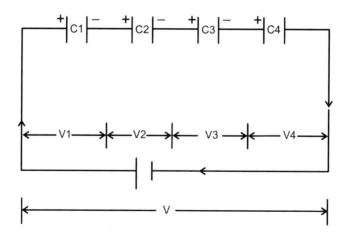

Fig. 13.4: Capacitors in series

- $1/C = 1/C_1 + 1/C_2 + 1/C_3 + 1/C_4$
- In general, $1/C = 1/C_1 + 1/C_2 + 1/C_3 + \text{---} + C_n$
- Where, n is number of capacitors in series.

Capacitors in parallel (Fig. 13.5)

If capacitors are connected in parallel, the total charge developed on them is sum of the charges on each of them.

- $Q = Q_1 + Q_2 + Q_3 + Q_4$ ....(3)

From formula (1)
- $C = Q/V$

From formula (3)
- $C = (Q_1+Q_2+Q_3+Q_4)/V$
- $C = Q_1/V + Q_2/V + Q_3/V + Q_4/V$
- $C = C_1 + C_2 + C_3 + C_4$
- In general, $C = C_1 + C_2 + C_3 + \text{---} + C_n$
- Where, n is number of capacitors in parallel.

Charging

Capacitor can be charged using electrostatic induction, where a static electric charge is allowed to build up on the plates of the capacitor, or by applying a potential difference across the plates from source of electricity.

Discharging

Capacitor can be discharged when accumulated charge is allowed to flow off the plates. If the two plates with opposite charges are connected, electrons flow from negative from positive plate until their charges are equal. The time taken for this discharge depends on the capacitance of the con-

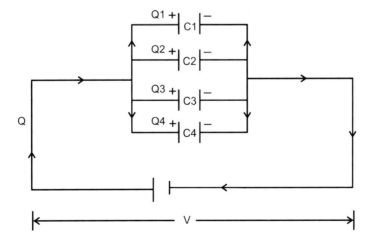

Fig.13.5: Capacitors in parallel

denser, the resistance (or inductance) of the pathway and quantity of electricity involved.

CAPACITIVE REACTANCE

Capacitive reactance is an opposition to the alteration of voltage across circuit.
- **Formula**
  - $X_C = 1/2fC$
  - Where, X_C is capacitive reactance (ohm)
  - f is frequency (hertz)
  - C is capacitance (farad).

MULTIPLE CHOICE QUESTIONS

1. Impedance offered by condenser to flow of current is called
   a) Inductive reactance
   b) Capacitive reactance
   c) Inductance
   d) Resistance
2. Capacitive reactance
   a) Increases with increased capacity of condenser
   b) Increases with increased frequency of current
   c) Is unaltered by the size of the capacitor
   d) Reduces with increased capacity of condenser
3. The duration of condenser discharge does not depend on...
   a) Capacitance of the capacitor
   b) Resistance of pathway
   c) Amount of electric charge
   d) None of the above
4. The dielectric medium usually employed are
   a) Mica
   b) Glass
   c) Plastics
   d) All of the above
5. The plates of condenser are normally made of
   a) Silver
   b) Aluminum
   c) Brass
   d) All of the above
6. If 3 capacitors of 10×10^{-6} farad, 20×10^{-6} farad and 30×10^{-6} farad are arranged in parallel, what will be resultant capacitance?
   a) 20×10^{-6} farad
   b) 60×10^{-6} farad
   c) 0.1833 mega farad
   d) 183300 farad

7. Select the correct one
   a) C = Q × V
   b) Q = C × V
   c) V = Q × C
   d) None of the above
8. One of the following is not affecting capacitance
   a) Size of the plates
   b) Material of the plates
   c) Color of the plates
   d) None of the above
9. Unit of capacitance is
   a) Farad
   b) Henry
   c) Watt
   d) Volt

Answers:

1—b	4—d	7—b
2—d	5—d	8—c
3—d	6—b	9—a

Bibliography:
1. Clayton's Electrotherapy. AITBS Publishers, 2002.
2. Singh J. Textbook of Electrotherapy, 1st ed. Jaypee Brothers Medical Publishers (P) Ltd. 2007.

Reference websites:
1. http://electronics.howstuffworks.com
2. http://en.wikipedia.org
3. http://hyperphysics.phy-astr.gsu.edu
4. http://whatis.techtarget.com
5. http://wiki.answers.com
6. http://www.britannica.com
7. http://www.ehow.com
8. http://www.thebigger.com

Physiology of Pain

14

International Association of Study of Pain has defined pain in these words, "Pain is an unpleasant sensory and emotional experience associated with actual or potential tissue damage, or described in terms of such damage."

Components of Pain

Pain has two components mainly:
- Fast pain
- Slow pain.

Comparison between fast pain and slow pain is given in Table 14.1.

Table 14.1: Comparison between fast pain and slow pain

Fast pain	Slow pain
Also known as first pain	Also known as second pain
That occurs during injury	That occurs after injury
It is carried by Aδ nerve fibers	It is carried by C nerve fibers
Intensity of pain is severe	Intensity of pain is less
Pain is localized at the site of injury	Pain is generalized around the site of injury
For example, when accident occurs and any bone is fractured at that time severe pain occurs	For example, once fracture has occurred, pain does not subside immediately. Even after fracture, dull aching pain persists for weeks

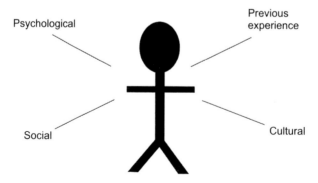

Fig. 14.1: Factors affecting pain perception

Physiology of Pain

Factors Affecting Pain Perception (Fig. 14.1)

Each and every person has different pain perception even if pain intensity and site are same.

Following are factors which will affect pain perception.

Psychological

In this world no two persons have same thinking. It means psychology differs from individual to individual. One, who is psychologically strong, will not feel pain much if he has locked his finger between two doors. But one, who is psychologically weak, will shout and may even cry if same happens to him.

Previous experience

Repeated experience of exposure to same pain stimuli will increase threshold of your particular pain perception. For example, if we touch hot silencer of bike, we will feel too much pain. But the person who is working in garage will not think also if he gets contact with hot silencer.

Social

Our family members, friends circle, colleagues also affect our pain perception. If we have simple knife injury, we will complain of pain and our attention will be continuously on that only. But army soldiers are living happily with bullet injury also because they are living in that type of society in their camp in which fear has no place to get attention.

Cultural

Our cultural background is also very important. Just for example, villagers will go bare feet everywhere and people living in city not going into kitchen without footwear. Villagers will not complain of pain due to thorn injury as frequently as city living people are doing as they are habituated with this type of lifestyle.

Pathway of Pain

Pain sensation from skin and deeper structures

It is carried by lateral spinothalamic tract explained later.

Pain sensation from viscera and face

Pain from thoracic and lumbar viscera is carried by thoracolumbar sympathetic nerves. And pain from esophagus, trachea and pharynx is by vagus and glossopharyngeal nerves.

Pain sensation from pelvic region

Pain from deeper structure is carried by sacral parasympathetic nerves.

The lateral spinothalamic tract is also known as lateral spinothalamic fasciculus. It is ascending pathway which carries sensations towards the brain. It carries pain and temperature sensory information to the thalamus of the brain. It is made up of myelinated A nerve fibers and unmyelinated C nerve fibers.

Brief Anatomy of lateral spinothalamic tract

- 1st order neuron: Free nerve endings are receptors for pain reception from body. Any cell damage will stimulate free nerve endings and impulse will start to flow in peripheral nerve fibers made up of Aδ or C nerve fibers. These sensory axons will enter into the spinal cord by posterior horn and synapse with 2nd order neurons.
- 2nd order neuron: Most of the second order neurons in spinal cord will cross to the opposite side and ascend up. Some of the fibers will ascend up on the same side without crossing midline. These fibers will ascend through the medulla and through the pons and midbrain and synapse in the thalamus, reticular formation and other structures.
- 3rd order neuron: The third order neurons from thalamus and reticular formation will then project through the internal capsule and corona radiata to various regions of the cortex, primarily the main somatosensory cortex, Brodmann's Areas 3, 1, 2. Because of its connection with reticular formation during pain perception sleep disturbance is common as reticular formation is concerned with arousal mechanism.

Center

Postcentral gyrus of parietal cortex is final destination for pain sensation.

Function

The fibers conduct impulses of pain and temperature. Sensation of pain and temperature will be felt only when impulse reaches at somatosensory cortex.

Neurotransmitter

Neurotransmitter for pain transmission is substance P.
For detailed anatomy please refer to Textbook of Anatomy.

Causes for Pain

- Chemical stimuli: During empty stomach, because of gastric juice pain will be felt.
- Spasm: After prolonged exertion, spasm of limbs will produce pain.
- Mechanical: After trauma feeling of pain is an example of mechanical pain.
- Ischemia: During ischemia of heart, pain is felt over chest.
- Polymodal: If more than one cause together are giving pain.

Referred Pain

Sometimes origin of pain is somewhere else, and pain will be felt on other region of the body. This is known as referred pain. Deep pain is referred but superficial pain is not referred. Examples of different referred pain are given in Table 14.2.

Table 14.2: Examples of different referred pain

Source organ of pain	Pain is felt over
Diaphragm	Right shoulder
Heart	Medial part left arm and left shoulder
Kidney	Loin
Gallbladder	Epigastria region
Ovary	Umbilicus
Testis	Pubic region

Dermatomal Rule

This rule is for referred pain. If two organs are developed from same dermatome during development in embryo, pain from one organ will be referred to another organ. For example nerve supply of diaphragm is phrenic nerve has root value C3, C4. Shoulder also has same root values of nerve supply so pain of diaphragm will be felt over right shoulder.

Pain Regulation

We know that pain will be perceived when it reaches to cerebral cortex. So it pain sensation is stopped in pathway only before it reaches to cerebral cortex, patient will not perceive pain. Its regulation occurs at following levels:

Peripheral level

Short wave diathermy, ultrasound, etc. will remove waster products from body part by increasing blood supply and will relieve pain. This is an example of pain regulation at peripheral level.

Spinal level

In spinal cord T cells are situated. These cells are responsible for passing all the sensation including pain. That is why they are also known as Wide Dynamic Range (WDR) cells. Now as per diagram large diameter fibers have facilitatory control over Substantia Gelatinosa (SG) and Substantia Gelatinosa has inhibitory control over T cells. These two structures we can compare with police and thief. When police is active, thief will be inactive and when police will be inactive, thief will be active. Same ways, when substantia gelatinosa is stimulated by large diameter Aβ fibers, T cells are inhibited. Due to inhibition of T cells, pain sensation will be stopped here and it does not reach to cerebral cortex because pain is transmitted by small diameter Aδ and C fibers. Ultimately patient will not feel pain. This theory is known as pain gate theory (Fig. 14.2).

Supraspinal level

Descending pain suppression system: There are two descending pain modulating systems. They will release some chemicals which are blocking pain transmission by various mechanisms. There chemicals are found in central nervous system.

One descending pain suppression system involves following neurotransmitters:
- Serotonin
- Dopamine
- Acetylcholine.

Above mentioned chemicals are relevant to action of antidepressants in relieving pain.

Second pain modulating system is mediated by neuromodulators like:
- Encephalin
- Beta-endorphin.

Above mentioned chemicals are working as natural painkillers because they have morphine like effects.

- **Counterirritation:** It also works at supraspinal level. When any electrotherapy treatment like transcutaneous electrical nerve stimulation, etc. or analgesic gel applied to skin, it will work as counter irritant by producing superficial sensation and will relieve deep pain.
- **Placebo:** It is very useful treatment in which actually treatment is not given actually but use of unfamiliar apparatus or drugs will make the patient that he has taken treatment and will start to feel pain relief. Like use of multivitamins it patients complaining of pain. It means when pain is psychological without any pathology, pain is in patient's mind. Patient will take any tablet and he will start to feel that now it is better.

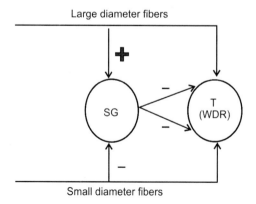

Fig. 14.2: Pain gate theory

Cortical level

At this level behavior modification, meditations, use of frankly atmosphere, laughter therapy will help the patient to forget pain even though pain is present.

Terminologies

- Allodynia: Pain perceived from nonnoxious stimuli (e.g. touch, pressure, etc.)
- Analgesia: Loss of pain
- Causalgia: Painful, burning sensation along the distribution of nerve
- Dysesthesia: Touch sensation is perceived as pain
- Hyperalgesia: Increased sensitivity to pain
- Hypoalgesia: Decreased sensitivity to pain.

MULTIPLE CHOICE QUESTIONS

1. One of the following sensation is unpleasant and emotional
    a) Touch
    b) Pain
    c) Temperature
    d) Pressure
2. One of the following is not affecting pain perception
    a) Social factor
    b) Previous experience of person
    c) Psychology of person
    d) Height and weight of person
3. Fast pain is carried by
    a) A α fibers
    b) A β fibers
    c) A λ fibers
    d) A δ fibers
4. Following are unmyelinated fibers
    a) A fibers
    b) B fibers
    c) C fibers
    d) None of the above
5. Pain after fracture of bone is example of which cause of pain
    a) Ischemia
    b) Chemical stimuli
    c) Spasm
    d) Mechanical
6. Referred pain of diaphragm is felt over
    a) Left side chest and arm
    b) Right side shoulder
    c) Epigastria
    d) Loin

7. Dermatomal rule is for following type of pain
    a) Mechanical pain
    b) First pain
    c) Second pain
    d) Referred pain
8. Select neurotransmitter responsible for pain
    a) Acetylcholine
    b) Substance p
    c) Dopamine
    d) Glutamate
9. Hyperalgesia means
    a) Loss of pain sensation
    b) Hypersensitivity to pain sensation
    c) Abnormal pain sensation
    d) Referred pain sensation
10. Pain gate theory works on which level?
    a) Spinal cord
    b) Brainstem
    c) Cerebrum
    d) Cerebellum

Answers:

1—b	4—c	7—d	10—a
2—d	5—d	8—b	
3—d	6—b	9—b	

Bibliography:
1. Sembulingam K, Sembulingam P. Essentials of Medical Physiology, 2nd ed. Jaypee Publication, New Delhi.

Reference website pages:
1. http://en.wikipedia.org
2. http://www.britannica.com
3. http://www.health24.com

Index

A
Absorption 69
Active components 1
Alternating current 25
Ammeter 11
Amplitude 26
Artificial magnets 57
Atomic theory of magnetism 59
Attenuation 70
Attenuation coefficient 71
Autotransformer 88

B
Bias voltage 35
Biogas 29
Biphasic pulsed current 25

C
Capacitance 95
Capacitive reactance 100
Capacitors
    in parallel 99
    in series 98
Causes for pain 104
Causes of electric shock 48
Cell
    hazard 34
Characteristics of electric lines of force 4
Capacitor 95
    charging 99
    discharging 99
    inventor of 95
    types of 97
Choke coil 38
Classification of current according to frequency 27
Classification of electrode 42
Comparison between electricity and magnetism 61
Components of pain 102
Conductors 16
Continuous alternating current 27
Cosine law 72
Coulomb's inverse-square law 9
Coulomb's law 9
Counter irritation 107
Current electricity 24
Cylindrical conductors 97

D
Decay time 26
Dermatomal rule 105
Descending pain suppression system 106
Diamagnetism 57
Dielectric 95
Diode 35
Direct current 25
Disposable cell 34

E
Earth shock 52
Earthing 31
Eddy current 82
Electric cell 34
Electric charge 2
Electric conductance 7
Electric current 3
Electric field 3
Electric field of capacitor 96
Electric potential energy 8
Electric resistance 5
Electric shock 48
    first aid 51
    medical uses 51
    prevention 51
    self-care 51
    signs and symptoms 49
    torture by 51
    treatment 51
    types of 48
Electric tension 4
Electrical energy 8
Electrical potential difference 4
Electrical power 8
Electricity generation 27
Electricity transmission and distribution 29
Electricity types 24
Electrochemical cells 34
Electrode gel 45
    characteristics of 46
Electrodes 42

Electrolytic cells 34
Electromagnetic induction
  inventor of 79
Electromagnetic spectrum 64
  history of 65
  types of 66
Electromagnets 58
Electromotive force 4
Electronics 1
Electrostatic potential energy 8
Energy loss from transformer 89
Equatorial line 57
Even ratio transformer 87
Extinction 70
Extrinsic semiconductor 17

F
Factors affecting attenuation 70
Factors affecting pain perception 103
Factors affecting skin resistance 41
Factors for electromagnetic induction 79
Factors in lethality of electric shock 50
Faraday's law 81
Fast pain 102
Ferromagnetism 57
Field effect transistors 92
Fixed capacitor 97
Forward bias 36
Foucault current 82
Frequency 27
Frequency 64
Fuel cells 34
Full-wave rectification 38
Fuse 21

G
Galvanic cell 34
Galvanic skin response 41
Geometric length 56
Geothermal energy 29
Grothus law 72
Grotthus Draper law 72
Grouping of capacitors 98
Grouping of resistance 6

H
Half-wave rectification 37
History of transistor 91
Hydropower 28

I
Inductive reactance 81
Insulators 16

Intensity 26
Interpulse interval 26
Interrupted alternating current 25
Intrinsic semiconductor 17
Inverse square law 69
Inverter 37

J
Joule's effect 9
Joule–Lenz law 9
Joule's law 9
Junction field effect transistor 92
Junction transistors 92

L
Lambert-Cosine law 72
Lateral spinothalamic tract 104
Law of refraction 68
Lenz's law 82

M
Macroshock 48
Magnet 56
  shapes 59
  types of 58
Magnetic effect of electric current 59
Magnetic flux 79
Magnetic length 57
Magnetic meridian 57
Magnitude 26
Mains supply 30
Metal electrode 43
Metal oxide semiconductor field effect transistor 92
Microshock 49
Molecular theory of magnetism 59
Monophasic pulsed current 25
Mutual induction 80

N
Natural magnets 57
NPN transistor 92
N-type semiconductor 18
Nuclear power 28

O
Ohm's law 6

P
Pathway of pain 103
Pain regulation 105

Index

Pain-gate theory 106
Parallel plate capacitors 97
Paramagnetism 57
Parameters of electrodes 44
Passive components 1
Permanent magnets 58
Phase 25
Phase duration 26
Photoelectric cells 34
Photovoltaic cells 34
Placebo 107
PNP transistor 92
Potential difference 4
Potential drop 4
Potentiometer 11
Power plugs and sockets 20
Precautions against earth shock 53
Primary cell 34
Properties of a magnet 60
Properties of magnetic lines of force 60
Psychogalvanic reflex 41
P-type semiconductor 17
Pulsatile alternating current 25
Pulse 25
Pulse duration 26
Pulsed alternating current 27

R
Rechargeable cells 34
Reciprocal engines 29
Rectifier 37
Referred pain 105
Reflection 67
Refraction 67
Renewable energy 28
Resistance in parallel 7
Resistance in series 6
Reverse bias 36
Rheostat
    variable resistance 10
Rise time 26
Rubber electrodes 42

S
Secondary cell 34
Self induction 80
Self-adhesive electrodes 44
Semiconductors 16
Series rheostat 10
Severity of shock 49

Shunt rheostat 11
Skin resistance 41
Slow pain 102
Snell–Descartes law 68
Snell's law 68
Solar cells 34
Solar power 28
Spherical conductors 97
Static electricity 24
Static transformer 86
Step down transformer 87
Step up transformer 87
Storage cell 34
Switches 21

T
Temporary magnets 58
Thermal power 28
Thermionic valves 35
Tidal wave energy 29
Transformer
    construction of 86
    principle of 86
    uses 86
Transistor
    history of 91
    types of 92
    uses 91
Triode 37

U
Uninterrupted alternating current 25
Uses of eddy current 83
Uses of electromagnetic waves 73
Uses of variable capacitor 98

V
Vacuum electrodes 44
Variable capacitor 98
Variable transformer 88
Velocity 65
Voltage 4
Voltaic cell 34
Voltmeter 12

W
Waveform 26
Wavelength 64
Wind power 28